HUMAN LEARNING: From Learning Curves to Learning Organizations

INTERNATIONAL SERIES IN
OPERATIONS RESEARCH & MANAGEMENT SCIENCE

Frederick S. Hillier, Series Editor
Stanford University

HUMAN LEARNING: From Learning Curves to Learning Organizations

by

Ezey M. Dar-El

Faculty of Industrial Engineering and Management
Technion – Israel Institute of Technology
Haifa 32000, Israel

Kluwer Academic Publishers
Boston/Dordrecht/London

Distributors for North, Central and South America:
Kluwer Academic Publishers
101 Philip Drive
Assinippi Park
Norwell, Massachusetts 02061 USA
Telephone (781) 871-6600
Fax (781) 871-6528
E-Mail <kluwer@wkap.com>

Distributors for all other countries:
Kluwer Academic Publishers Group
Distribution Centre
Post Office Box 322
3300 AH Dordrecht, THE NETHERLANDS
Telephone 31 78 6392 392
Fax 31 78 6546 474
E-Mail <orderdept@wkap.nl>

 Electronic Services <http://www.wkap.nl>

Library of Congress Cataloging-in-Publication Data
Dar-El, E.
    Human learning: from learning curves to learning organizations / by Ezey M. Dar-El.
        p.cm -- (International series in operations research & management science; 29)
    Includes bibliographical references and index.
    ISBN 0-7923-7943-8
    1. Adult learning. 2. Organizational learning. I.Title. II.Series.

LC5225.L42 D36 2000
153.1'58--dc21   00-058398

Printed on acid-free paper.

Printed in the United States of America

To my wife
Evelyn

CONTENTS

Preface

Human learning is a topic that affects virtually every human endeavor ever undertaken. No matter what the activity, we call on previous (or similar) experiences and knowledge to help us plan and predict the 'best' manner to proceed with its execution. This is especially important each time an activity is repeated; here, we bring into play all our previous experience and knowledge of the specific processes involved.

Motivation for writing this book developed when it became evident that only a few books were written on this subject; and what was written, merely covered the essentials of fitting learning curve formulae to generated field or laboratory data. This is not the purpose of 'learning curves'. These are needed for use in a predictive sense, in order that one can cost and plan the activities and resources needed for accomplishing some task.

Considering the practical applications for using learning curve theory, we come across many blanks in our knowledge, and the best we are able to deal with are the expected learning characteristics using completely *naive* operators working on fairly straightforward (mainly) motor tasks.

However, even determining 'best estimates' of the two learning parameters for the simple power model case is problematical. In fact, except for one paper, nothing is published for estimating beforehand (i.e., without experience) the execution time for the first cycle, and furthermore, learning slopes can only be 'guessed at' from what appears to be similar tasks found in Learning Slope tables of previous studies.

But what is more important, there is no way to account for previous experience, nor do we have too many clues as to how to handle the forgetting phenomenon (which must necessarily apply in every conceivable maintenance situation).

So - the decision was made to write a book on Human Learning with the intention to fill in the 'gray' areas, as far as possible. And there are plenty of 'gray' areas – so much so, I included a final chapter that summarizes the many topics that have yet to be properly researched.

After much hesitation, I decided to include a chapter on Learning Organizations, since I believe this topic has its roots firmly set in individual learning, and therefore should be included.

The book is designed to support classwork in two general study areas – Industrial Engineering and in Business Administration and/or Management. As a rule, courses on Human Learning/Learning Curves/Learning Organizations simply do not exist, though these topics do enter into many courses that are taught at both undergraduate and graduate levels. The obvious ones are concerned with Work Study and Operations Management. The less obvious ones deal with 'Productivity and Quality' management courses, Pay and Compensation and in Organization Behavior and Development. Maybe there are others as well.

ACKNOWLEDGEMENT

I wish to thank the Technion and the Dean of the Industrial Engineering & Management Faculty for supporting my sabbatical leave which made it possible for me to undertake the writing of this book.

My hat goes off to Dr. Shimon Nof who recommended the present name for the book, and for making suggestions on inclusions. The same goes for the suggestions made by the other 'blind' referees. I thank my good friend, Jorge Yamamoto, a fine industrial psychologist, who reviewed and commented on Chapter 9, on Learning Organizations. Jorge insisted that I distinguish between 'Learning Organizations' and 'Organizational Learning', although, even today, most writers on this topic do not distinguish between the two!

My acknowledgements would be incomplete with the wonderful assistance I received from the staff of our Industrial Engineering & Management Branch Library, but specifically Tatyana Shraiber and Jody Bar-On, who kept feeding me with information and copies of journal articles.

Next in line for thanks are the Technion's IE&M Faculty's 'English' secretaries, Eva Gaster and Lillian Bluestein, but particularly the latter, who always smiles and says "it's no problem" as I piled on the work! Finally, my acknowledgements go to Gary Folven of Kluwer, who gave me the support when it was most needed, to complete the book.

1 INTRODUCTION TO HUMAN LEARNING

We gain experience each time we repeat some activity. The experience takes the form of either or both improved manipulative skills and in process procedures. This is especially important during the early 'cycles' (or, 'repetitions'), when each repetition adds both confidence and usually, improved speed in accomplishing the performed task. As the number of repetitions increases into the hundreds, improvements in learning are not so discernable between adjacent cycles. Rather, improvements are only seen after large numbers of additional cycles are completed.

We call this characteristic the "Learning" phenomenon. We cannot exactly explain why learning occurs, except that it does. Learning is also referred to as 'experience', and indeed, when we refer to a person as being 'experienced', e.g., as with a consultant, we assume this person has successfully undergone these activities (or ones of a similar nature) on several occasions. The plotted relationship between the time taken to complete some activity 'n' times, is called the 'Learning Curve', or, 'Experience Curve'.

The term 'Learning Curve' is used for describing labor learning at the level of the individual employee, or, the production process, such as an assembly line. But we need to distinguish this type of learning from improvements that occur at the process and plant level, right up to the level of the entire firm. We shall refer to this type of learning as "Progress Functions", a term commonly used in the literature for learning at the firm and industrial levels.

Early Learning Curve research was an outcome of pioneering investigations done by experimental psychologists who studied reaction times and percentage recall of simple tasks, such as pressing a sequence of buttons on cue, as well as more complex tasks, such as recalling memorized poetry versus an equivalent number of non-sensible words; the latter, in order to demonstrate that recall was a lot more effective with poetry than it was with learning non-sensible words.

The characteristics of the learnt activity have a major influence on the nature of the learning process. Some activities can be very simple, such as assembling a 4-piece jigsaw puzzle, where learning is very rapid, but the potential for improvements are quite limited. Others can be far more complex. For instance, to correctly swim the crawl stroke is a very complex task, and in most instances, requires a great deal of practice. This is also true for the exact assembly of a guided missile. Performance times may appear to be painfully long at the beginning, but the potential for improvement can be very large.

Bryan and Haster published an article in 1899, on the acquisition of skills in relation to the learning of Morse code by telegraph workers. Cochran (1969) quoted a passage from the early work of Taylor (at the turn of the century) referring to allowances given during the learning period: "Study and record the percentage which must be added to cover the newness of a good workman for a job, for the first few times he does it. This percentage is quite large on jobs made up of a large number of different elements composing a long sequence, infrequently repeated. This factor grows smaller, however, as the work consists of a smaller number of different elements in a sequence that is more frequently repeated."

As mentioned before, the earliest learning studies were done by experimental psychologists (see Ebbinghaus, 1913; Snoddy, 1926), but the most important learning theory work and industrial application was done by Wright (1936) on aircraft production. Despite Wright's model dealing with a very large product (and therefore involving a lot of "organizational learning"), an extraordinary number of researchers applied his model to individual learning situations – and with fairly good results (see for example, the work by Conway and Schultz, 1959; Konz, 1983; Baloff, 1966, 1971). Wright's learning Curve model is generally referred to as the "Power Curve", or the "Power Model".

More recently, psychologists, including Newell and Rosenbloom (1981) and Goonetlleke et al. (1995), concluded that the power curve was also applicable for a wide range of human tasks including "mechanical behavior, perceptual-mechanical tasks, perceptual tasks, elementary decisions, memory tasks, complex routine tasks, and problem-solving tasks (Arzi and Shtub, 1997)".

Industrial engineers almost always use the power model to describe learning without regard to the operator's characteristics, or the circumstances in which the work is applied.

Learning has also been characterized by whether the repetitive work is performed by individuals, or by groups of persons. Thus, we differentiate between individual learning and organizational learning. The latter can be further dichotomized as "product learning" and "organizational systems learning"; the former dealing with the learning of large products (e.g., aircraft manufacture), while the latter involves the actual functions of the organization and is not necessarily associated with a specific product. The

Learning Organization is currently a very 'hot' topic and much time and effort is being devoted to its research and development.

Individual learning, group learning, product learning, organizational learning and learning organizations are all elements of Human Learning, the title of this book.

Learning curves were first used in the aircraft industry until the Second World War. Later on, their use spread to other industries, such as the shipping, oil and automotive industries. The outcome of applying learning theory is very clear – a firm can expect continuous improvements of its productivity ratios as a consequence of increasing its experience, or, 'stock of knowledge'.

Research on the acquisition of skills (Barnes and Amrine, 1942) can be separated into two groups: the most extensive one deals with the behavioral scientists (see for example, Van Cott and Kinkade, 1972; Anzai and Simon, 1979; and Mazur and Hastie, 1978). Their research methods consisted mainly of laboratory experiments dealing with the responses to various stimuli. One element was always present in the tasks being studied – the element of decision-making. Some decisions about the next movement need to be made. The tasks varied from simple ones, such as pressing a button when a light comes on, to more difficult tasks where the number of lights and buttons are increased, then finally to extremely difficult and complex tasks such as pilot training. In the simple case, there may only be one stimulus, e.g., one light with one possible response – one button to press. In the latter cases, the number of stimuli and possible responses are increased, so that the decision may be when and what button to press. In the case of pilot training, the trainee must choose the best set of responses for any possible set of stimuli. Here, the number and range of stimuli-response combinations is so large that it is virtually impossible to enumerate every possible one.

Several hypotheses have been proposed as to how industrial workers learn. Carlson (1962), DeJong (1964) and Hancock et al. (1963) found there was a decrease from cycle to cycle in the number of times the eyes have to be refocused. Discontinuous and hesitant movements become smoother, fumbling disappears, and greater simultaneity in movements is perceived. Crossman (1959) says that an unskilled worker faced with a new task tries out various methods and retains the more successful ones. He adds that expert ability seems to lie in knowing exactly the right method to use in each situation that arises, rather than having superior coordination, acuity or timing. The experienced worker is able to select the right source of signals to attend to, choose the right course of action, make precisely the right movements and check the results by the most reliable means. Caspari (1972) performed an MTM analysis of a task at different stages of learning. He found that as experience was gained, the worker uses fewer MTM movements to perform the task. Gershoni (1979) also found that workers were able to reach standard velocity according to MTM standards even at the start of the learning process. This confirms Crossman's claim that progress is achieved mainly through more efficient patterns and not through speed of motions. Dudley (1968) showed that a learner's distribution of performance times is symmetrical, but with practice, becomes positively skewed. However, the actual range of the cycle does not change.

The need for good models on Learning Curve Theory can be best understood via its applications. For individual learning, the obvious needs are: for determining manpower policies, for Labor Costing (Smith, 1989); for Production Planning; for setting Time Standards; for establishing Wage Incentives; for planning Batch Production lot sizes; for determining the minimum number of Assembly Stations; for determining Optimal Cycle Times for assembly; and so it goes on and on.

For large products (such as aircraft, ships, buildings, etc), the major concern is Costing – to acquire a competitive budget and then to check actual times against the budgeted times. Other obvious applications are Production Planning and Resource Planning, and Control for project environments (see Richardson, 1978). Figure 1.1 was used by Rosenwasser (1982) in his MSc thesis which shows that learning curves play an important role in many individual and related functions in production systems.

However, these are the more obvious uses of Human Learning. The less obvious but more dramatic uses are in the results of work teams in tackling organizational learning issues. Continuous Improvement teams, concurrent engineering teams, Quality Circle teams, Work Improvement teams, and so on, have all made their contribution in human learning, resulting in productivity improvements unmatched by individual learning. We have seen, in the last decade and more, a focus on organizational improvement systems, such as JIT (Just-In-Time), TQM (Total Quality Management, and Reengineering, just to name a few.

The 'economists' began to develop models for price elasticity and to determine the conditions under which a firm (or, even an industry) could utilize a mathematically developed policy for optimizing profits. With the introduction of rapid design changes, we see product costs continually dropping. It should not come as a surprise that Learning Curves are used for predicting the cost of future products belonging to the family.

We can say that Human Learning has developed almost independently, in four areas. The first deals with individual learning and includes many factors such as forgetting and relearning. Forgetting is just emerging as a research area. It occurs everywhere, and we have only just begun to scratch the surface in understanding human behavior in the learning-forgetting-relearning cycle. The second deals with product learning where the focus is on integrating the efforts of individual learning together with improvements

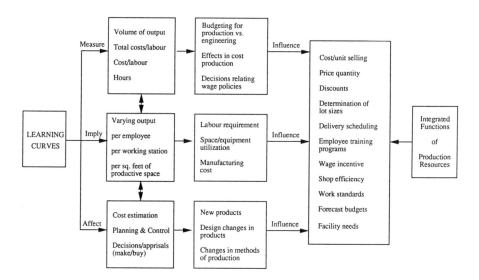

Figure 1.1: Integration of Learning Curves and Production Resources

in all aspects of materials and information flow involved in putting the whole product together. It deals with manufactured product learning and provides important inputs in determining whole product costs.

The third area deals with rapid changes in product development, e.g., TVs, videos, PCs and so on. We're talking about product design changes as the essential ingredient in this type of learning, and at times, with an emphasis on design 'quality'. Here, human learning focuses on introducing design and manufacturing process changes, so that the learning curve of such products is associated with very high learning slopes (around 60%).

The fourth area deals with Learning Organizations in its broadest sense, since it incorporates virtually every function of the organization. The system gradually undergoes a cultural change, learning and improving as it involves itself in the change process. This is the ultimate in Human Learning. Few organizations are able to operate at this level, but we must learn what these concepts are all about and begin to ask how we can move towards its achievement.

This book essentially focuses on Human Learning at the individual level – simply because this is fundamental to the other three areas of learning as well. Thus, Chapter 2 deals with the factors that influence learning curves for individuals. The main learning curve models for individuals without forgetting are covered in Chapter 3. There are worked examples following each model, so that the reader will be able to follow the model applications.

Parameter estimation is covered in Chapter 4, where the emphasis is on predicting learning curve performance without having any previous experience with the job on hand. This is by far one of the most daunting tasks facing Human Learning. We need to model learning behavior, but we need to do this during the planning stages in order to plan its production. We cannot assume a layout of the task in order to find the best layout for its manufacture!

Chapter 5 deals with learning models with forgetting. Not much is known about 'forgetting', but this chapter makes the best of what is available. Applications are given in Chapter 6. We try to include every area of application available on Human learning at the individual and planning levels. The reader should find the need for developing learning curve applications via the experience of those who have reported their investigations in the literature.

Chapter 7 deals with cost models. These are specifically aimed at finding the optimal (re)training programs for employees to undertake in order to ensure that they remain in a high level of training during the period of their special duties, especially in situations where errors can result in heavy financial losses, and/or loss of life. We refer here to monitoring duties in the control of large plants – nuclear, power generation, chemical, etc.; crane operators, passenger bus and train drivers, ship personnel, etc.

Chapter 8 deals with Product Learning as an extension of individual learning concepts. Included will be product learning subject to rapid design and process changes, as well as an introduction to econometric applications. Learning Organizations will be covered in Chapter 9 and the final Chapter 10 discusses future needs of Human Learning.

2 FACTORS THAT INFLUENCE THE LEARNING CURVE

Before we enter into discussions on the various factors that influence learning curve theory, there are certain conditions that must be met in order to permit one to know that learning actually takes place. The first is concerned with management's ability to plan, implement and control activities of the organization and the second is its ability to know what to measure, maintain standards and to properly document and maintain all relevant information regarding on-going production.

These would be the same requirements needed if one were planning to install an incentive plan in a plant. The actual performance measure can vary, being dependent on the purpose of the application. It could be: direct labor hours; error rates; quantity produced per hour; setup times; maintenance time for a frequently occurring activity, and so on.

This is especially important for all 'start-up' conditions in a company's recent past and is of special importance to graduate students who, if fortunate, are able to flush out reliable data for learning curve analysis (as was the case with Rosenwasser, 1982).

Conway and Schultz (1959) do a superb job in analyzing pre-production factors. These include: tooling; equipment and tool selection; product design; methods; process design; testing; inspection and shop organization. During learning itself, subsequent changes take place including: tooling; methods; design; management; volume; quality; wage plans and operator learning. Conway and Schultz question whether the improvements measured are due to "learning" when a host of other factors are operating simultaneously to improve performances. However, we have no choice but to call this "learning", but we must understand that the source of the improvements come from other than the acquisition of skills.

The factors that influence learning are listed below together with their appropriate section numbers. The order in which the factors appear bears no relationship to their importance. The factors are:

2.1. Methods Improvement

Methods improvement is an essential step to take before the full benefit of learning can be realized. Experienced engineers need to analyze the methods by which the task is done in order to determine an efficient way for its execution. It is also an essential step to take if it is intended that the operator be put on a wage incentive plan.

Learning data derived from situations in which methods improvement was not applied simply cannot be relied upon to reflect true learning conditions. It is far too possible for an operator to begin work with a faulty method, and then to gradually introduce subtle changes in methods which can significantly reduce the performance time. The improvement in performance time is not due to learning, but to method changes which could have been avoided in the first instance.

This doesn't mean that a process cannot be improved beyond what is determined by the methods engineer. The latter simply tries to apply the principles of 'time and motion study' in order to define an efficient method to execute the task (see for example, Barnes, 1980; Mundel and Danner, 1995; Konz, 1995; and Kargar and Bayah, 1977). The operator, apart from acquiring the manipulative skills to accomplish the work more efficiently, may also introduce special jigs, fixtures, or even make sequence changes to improve performance times, usually in order to increase incentive earnings.

2.2. Worker Selection

It goes without saying that workers should be selected who are judged suitable for the job. 'Suitability' implies that they have both the minimum educational requirements as well as vocational training. Next, such workers need to undergo tests to ensure their ability to undertake tasks, or, if need be, to perform cognitive tasks. There are many worker selection tests in use, and one need only check with any library to find a rich literature on this subject.

With 'suitability' out of the way, the next factors that could influence learning are individual differences both between and within each worker.

One)   Differences between workers

Natural differences occur between individuals who have similar educational and vocational training, and previous work experience, over a large number of factors. These factors include, lung capacity, arm strength, natural walking speed, manipulative skills, just to name a few.

For example, consider a college group of, say, 100 male students of about the same age. Place them on a running track and instruct them to walk for an hour with 'purpose'. See that the percentage difference between the 'best' and the 'poorest' will not be too far from "2". Do the same test with lifting weights, or, with arm strengths. These differences reflect the natural differences that occur between workers.

When we omit 'pathological' cases from both ends of a large distribution, industrial engineers, for years have accepted that the natural ratio between the 'best' and the 'worst' is in the order of '2:1' (for example, see Barnes, 1980). Of course, we are referring to the ratio expected from a very large population. If we only use a few subjects in our experiments, then we could expect the ratio for any particular factor to be much smaller.

This variability must also occur in learning. Thus, experimental studies with learning *must* involve several subjects in order to obtain average values for parameter estimation. Pooling the work of several subjects would generate a broad band learning curve similar to that seen in Figure 2.1.

Two)   Variations within each operator

Even the variability within each person for the first, and sometimes up to the fourth cycle is quite large. Thus, it would be totally incorrect to use the first measured value of the performance time as the estimate of the parameter, t_1, the performance time for the first cycle (see Conway and Schultz, 1959; and Globerson and Gold, 1997). This variability is illustrated in Figure 2.2, which is typical of time outcomes for several operators doing the same task, over the first few cycles. We therefore need to take data from several cycles and use mathematical regression to extract the performance for the first cycle.

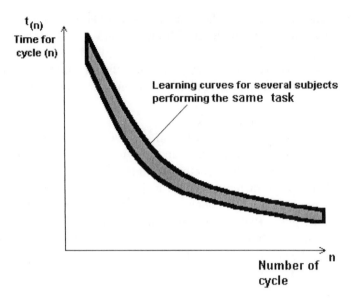

Figure 2.1: Illustrating the 'Between Operator' variability

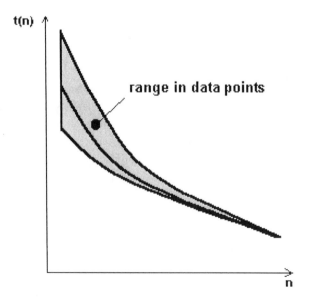

Figure 2.2: Illustrating the 'Within Operator' variability

2.3. Previous Experience

What constitutes 'previous experience'? Is work on a *similar* task to be considered as previous experience? If "yes", should we consider the experience to have a '1 to 1' correspondence? For example, a person has several years' experience working on the external bodywork of several products. Should he now be considered as having previous experience when he is scheduled to do the bodywork for a new product? The answer is a definite "Yes", but the question is, *how much* 'experience' should we say he has?

We can illustrate this situation in Figure 2.3. Curve A is what we'd expect a naive worker to follow. If we can conclude that a person's previous experience is worth 50 cycles in the new product, then our operator would follow curve B. If his experience is worth 150 cycles with the new product, his learning curve would be curve C.

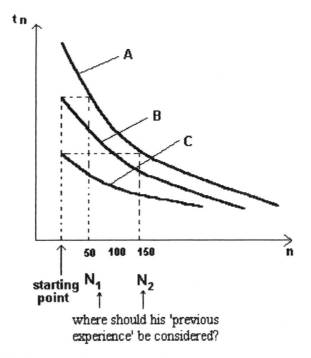

Figure 2.3: The case of previous experience

The matter is even more complicated when an operator acquires excellent skills on one type of work, but is then asked to work on a different type of work. Could he be considered to be naive? If 'No', then how much 'experience' should he be considered to have?

The effect of a transfer of skills was reported by several researchers. Von Tetra and Smith (1952) found transfer effects both in 'travel' and manipulation movements. Crossman (1959), who says that skill consists mainly of selecting the correct method for each situation, claimed that the transfer of skill from one task to another will take place when methods used for the old task are also appropriate to the new one. However, the amount of skill transfer will depend more on which criteria are selected than on the mere coincidence of methods between the tasks. Carlson and Rowe (1976) found that the initial learning is a function of the amount and proximity of the previous experience. Hoffman (1968) proposed a method for calculating the influence of prior experience on learning curve parameters. But the method requires an evaluation of previous experience and how much of that experience was forgotten. Lippert (1976) concluded that conventional methods of determining mean performance times without assessing prior experience does not result in serious errors.

Although the existence of knowledge transfer has been recognized, no criteria have been developed to evaluate its effect when an experienced worker changes from one task to another. For example, consider how long it would take two workers, having different degrees of skill, to perform the same task. At the moment, we have no idea on how to tackle this question.

Very little has been done to quantify 'experience'. Indeed, virtually all researchers go out of the way to ensure that their subjects have **NO** previous experience with the task to be studied. The reason is simple: subjects are supposed to clearly demonstrate the rapid learning associated with the lack of experience, and secondly, it is enormously difficult to assess how much experience to allow for 'experienced' subjects! But to do any effective work on prediction, this factor MUST be considered. This should be a challenge for future researchers to consider.

2.4. Training

Training is the stuff through which champions are made. There is no sports person who does not spend an enormous amount of time on training. Coaches try to identify the specific movements that need to be worked upon in order that their protégés achieve that little extra in their respective performances. It is no different in industry. To obtain superior performance, we need to identify the specific elements of the task that are likely to prove difficult to execute and train the worker in those specific activities until actions are performed speedily and without error. Monitoring the workers should continue until the task is performed flawlessly at high speed. This is a common method used for training new workers in textile mills that are heavily capitalized and whose machines operate at high speeds. Placing an untrained worker into such a mill could prove to be disastrous, since corrective work is needed at high frequencies and it is unlikely that the new worker would be able to cope (e.g., tying snapped threads on a bobbin). New workers are therefore trained outside the production halls on identical equipment until they are able to work at acceptable speeds without error.

Another example with textile factories was experienced by this writer. It was common practice to train new workers over three weeks in a production hall designed specifically for training new workers. Those who graduated would then be permitted to enter the real plant to begin their regular work. A European consulting firm specializing in textile production, introduced a training method similar to that described earlier, and succeeded in reducing the off-plant training time to just one week (Werner, International, 1972). The savings in reducing training time are indicated by the shaded portion in Figure 2.4. Eventually, whatever training method is used, we can expect the performance times to merge to an asymptotic value – but the potential for saving can be substantial. The essential mechanisms used by these consultants were "feedback" and "selection". The work of each trainee was displayed on a performance chart drawn with the average expected performance graph together with a lower tolerance line of "1 SD" (one standard deviation of the process) drawn in parallel to the average graph. The trainee was dismissed if, by the end of the first day, with intensive training, his/her performance was *below* the "1 SD" line. Such control charts are used very effectively as a performance motivator (see also Glover, 1966).

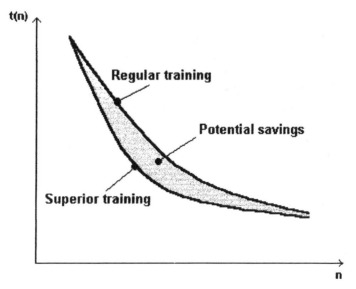

Figure 2.4: Savings potential via superior training

2.5. Motivation

The message is clear: "no motivation – limited learning"! Why should a worker put himself/herself out to achieve superior performance, if there is no perceived payoff? We can even stress the opposite and state, the stronger the motivating environment, the faster the learning. In Israel, during the Yom Kippur war in 1972, factory personnel in some plants were reduced by as much as 80% (because of the call-up for military duty). Yet for the approximate three-week duration of the war, factory output was barely compromised. While this has nothing to do with learning per se, the situation illustrates the power of motivation.

With most instances in industry, motivation takes the form of financial payments (such as, wage incentives). However, there are other motivators that operate and 'work system' designers need to exploit every possible avenue for creating the best possible environment for learning and for work. For example, "free time" is an even stronger motivator than is money rewards (Dar-El, 1980). When workers were told they could hold a Christmas party only after they had completed the day's production, imagine how the General Manager felt when the entire day's work was completed by late morning (Dar-El, 1960). Workers are fully conversant with the learning curve and Conway and Schultz (1959) and Mitchell Fein (1973, 1981), have written about workers holding back their work until *after* the incentive times are set.

In many experimental settings, subjects (usually students) are generally paid a nominal monetary incentive at the conclusion of their work, and researchers refer to this as a motivating environment (see Gershoni, 1971; Globerson, 1980, 1984; Dar-El et al., 1995). In other experimental settings, when students are obliged to do the work as part of a class exercise (Gilad, 1998), or, as determined by a military commander (Vollichman, 1993), no financial rewards are necessary since there is a high motivation to get away from the experiments as quickly as possible. Free time is an excellent motivator, and used as a reward, can provide a very good learning environment.

However, can "free time" be used as a motivator for regular work? (see Dar-El, 1980).

2.6. Job Complexity

Job complexity is highly correlated with the amount of information needed to accomplish the task. Indeed, some researchers define Job Complexity as the amount of "Information" needed for the task (see Fitts and Posner, 1967; Kvalseth, 1976, 1978; and Rosenwasser, 1982). The greater the information content needed, the longer it would take to completely learn the task, the sharper the learning rate, and the greater the potential for improvement. The lower the job content, the lower the information content, the easier it is to learn the task and the lower the learning rate.

The concept of information content is easily illustrated with a pack of shuffled cards. We need more information to sort the pack into its four suits (types) than we need to sort them into the color suits (i.e., 'blacks' and 'reds'). Hence, we are safe in concluding that it would take longer to sort into the four suits than it would to do the 2-colour separation.

As it is, 'job complexity' has a very strong influence on the learning rate, 'b', and to a lesser extent on 't_1', the time needed to perform the task for the first time. Both "b" and "t_1" are the major parameters needed to be evaluated for any mathematical model based on the power model. It is in this context we note that "b" can vary considerably from job to job (see Hirschman, 1964; Baloff, 1971; and Hancock and Bayah, 1992). This aspect will be discussed further in Chapter 4, which deals with parameter prediction.

2.7. Number of Repetitions (or cycles)

Learning curve performance times, typically 'flatten' out as the number of cycles reach large numbers (see Conway and Schultz, 1959). The Learning Slope is defined as the percent the performance is reduced by, each time experience is doubled. Note, this only applies for the power curve whose learning rate remains fixed during all production (see also Salvendy and Pilitsis, 1974). Almost all textbooks refer to a learning slope of 80% as being the 'most typical' value found. This is totally misleading, since we now know that learning slopes for individual learning may vary between 70% to 90% (Dar-El et al., 1995), and possibly even wider ranges, from 65% to 95%.

Table 2.1 illustrates the expected performance times for three power model learning slopes – 70%, 80% and 90% respectively. Starting times for each learning slope is assumed fixed at 100 time units, and each row shows the expected times when production is doubled.

We observe that the major reductions in performance times occur during the early part of the learning process. For example, about a one-third reduction in performance time is achieved in: 2 cycles (70%), 4 cycles (80%) and 15 cycles (90%) respectively. The next third reduction (i.e., a two-thirds reduction altogether) requires an additional 6 cycles (70%), 28 cycles (80%) and 1009 cycles (90%) respectively.

The second observation we make is that cycle times for both 70% and 80% learning curves become 'unrealistically' small, as the experience runs into the early thousands. This is the weakness of the power curve. Performance values drop to zero as the number of cycles increases – which means the power curve makes a poor fit for large numbers of cycles. This characteristic has been known for many years and Chapter 3 includes models that counter this 'zeroing' effect.

Table 2.1 Showing the Effect of Varying Learning Slopes

Number of cycles	Learning slope 70%	Learning slope 80%	Learning slope 90%
1	100	100	100
2	70	80	90
4	49	64	81
8	34.3	51.2	72.9
16	24.0	41.0	65.6
32	16.8	32.8	59.0
64	11.8	26.2	53.1
128	8.3	21.0	47.8
256	5.8	16.8	43.0
512	4.1	13.4	38.7
1024	2.9	10.7	34.8
2048	2.0	8.6	31.3
4096	1.4	6.9	28.2

2.7.1 Does learning continue forever? The most criticized shortcoming of the power model is that it does not account for any leveling in the production rate, i.e., it does not forecast any 'steady state'. Smith and Smith (1955) found a plateau appearing late in the learning curve which they claim is a physiological limit primarily related to manipulative movements, but other factors come into play (see Conway and Schultz, 1959). Baloff (1971), who uses the power model in his investigations, said that a prior assumption of continuous learning is risky in industry. He foresees the occurrence of a steady state phase, but cannot predict when it will occur. In any case, his assumption was to separate the learning phase from the steady state phase and treat them separately in practice. In some models, the existence of a steady state is incorporated within the model. DeJong (1957) expressed it as a factor of incompressibility, which takes into account not only the machine time, but also considers specific characteristics of the task.

However, the more popular approach to overcome the power model deficiency, is to use a continuous exponential model which forces the function (production, or, time) to asymptote (see papers by Bevis et al., 1970; Glover, 1965; Levy, 1965; and Pegels, 1969).

According to Kaloo and Towill (1979), the learning phase can extend from a few weeks to a year depending on several factors, such as, degree of automation and task complexity. There seems to be an underlying agreement among the researchers in accepting that workers achieve a steady state in their learning curves. On the other hand, when conditions for organizational learning occur, they may never ever reach a steady state (e.g., JIT methodology).

There are two references in the literature which state that learning continues even in the millions (see Crossman, 1959; and Glover, 1966). However, in practice, when the number of cycles runs into the thousands, performance times tend to asymptote at some ratio 'f' of the MTM (Methods-Time-Measurement) time, where f ≈ 0.80. This level usually represents the maximum incentive earnings the operator is prepared to aim for. The lesson is burnt deep in the operator's psyche, that when incentive earnings appear 'too high' to management, steps are taken through its industrial engineers to find ways to tighten the time standards (Fein, 1981).

Thus, we can say that workers determine outputs and it is in their interest to limit the output that will enable them to receive an acceptable incentive earning. This then, is my explanation for the 'sudden' appearance of the steady-state condition referred to by Baloff (1971) and Conway and Schultz (1959).

2.7.2 How many cycles to reach Time Standard? The question is, after how many cycles does it take for a worker (on average) to reach the Time Standard? Kilbridge, (1962) found that the number of cycles was an increasing function of the Standard Time. Hancock and Foulke (1963) found that for very short tasks (0.05 to 0.10 min.), more than half of the trainees reached the MTM Standard (Time) after 200 cycles, and all of them, after 900 cycles. Carlson and Rowe (1976) found that for tasks of about 10.0 min., 5000 cycles were needed to attain the Standard Time. These results confirm Kilbridge's finding on the number of cycles needed to meet the Time Standard, being a function of the task length. But research thus far is unable to find a simple relationship for the number of cycles needed to achieve the MTM task time.

2.7.3 The size of job orders. Today, who still works manually in the 'millions'? The real need for Learning Curve Theory is to come up with answers for the manufacture of very complex products needed in the 10s and 100s• but rarely beyond. The life of products are continually being shortened, with updated products in some industries (notably in the electronic industry) being superseded by still newer ones before customers have had a chance to absorb the previous model. Today's emphasis is on the rapid introduction of *new* products with fast but limited sales in each.

Thus, our major interest is in what happens during the early cycles when the largest productivity improvements are achieved. Remember also, that previous experience on product updates cannot be ignored.

This is where Human Learning needs to concentrate.

2.8. Length of the Task

The longer the basic time of the task (e.g., its MTM time), the longer it would appear to take a person to learn and the larger would be the value of 't_1', the performance of the first cycle. But this may not be strictly correct, when we consider that the task may contain several repetitions of sub-tasks. A sub-task is defined as "...a distinct describable and measurable subdivision of a task, and may consist of one or several fundamental motions" (Karger and Bayha, 1966).

Figure 2.5 is drawn to show that the overall task is comprised of a number of sub-tasks, some of which are repeats. Thus, the overall task includes, 2 sub-tasks A, 3 sub-tasks B, 3 sub-tasks C and a single sub-task D. Each time the whole cycle is completed, the operator does sub-task A twice, and each of sub-tasks B & C, three times, so that 'internal' learning occurs within each cycle. How would this affect the performance time for the first cycle?

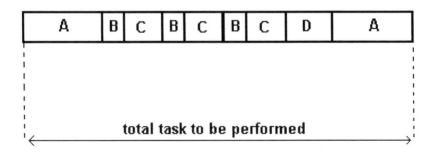

Figure 2.5: Breakdown of overall task into subunits

Globerson and Crossman (1976) add the times of unique sub-tasks to give the NRT (Non-Repetitive Time) of the overall task. Thus, the NRT for the task in Figure 2.5 would be the basic times for "A+B+C+D", suggesting that the learning parameters may be influenced more by the NRT values than by the overall MTM value.

We can extend this argument by claiming that all work is merely a combination of a limited number of basic micromotion elements defined at the MTM-I level, and therefore the equivalent NRT would be the sum of unique micromotion elements that appear in the overall task. This significantly narrows the expected values for 't_1', since the number of micromotion elements are limited – and this simply does not fit reality in any way.

The aspect of element sequence has not received much attention in the literature. Taylor pointed out that allowances for new workers would be smaller if some sequences of elements were repeated more frequently. Cochran (1969) pointed out that in cases where the final product contained several sub-assemblies that were very similar or equal, then for each final product completed, would mean more than one sub-assembly would be produced. Thus, if the product were an airplane, two wings would be produced for each completed airplane, so that the learning for the wing production would occur faster than for the aircraft itself.

Hancock et al. (1965) explained the lack of fit of actual data to the predetermined learning curve by the fact that the task performed contained many motions which were exact replications of each other, and learning took place at a faster rate than would normally be predicted.

Later, Hancock (1971) altered the model in order to include the effects of similar sub-cycles within the task.

In another report, Chaffin and Hancock (1967) defined a complexity index which took into account the repetition of segments within a task. Two segments were deemed to be identical when at least two sequential motions were the same in two different parts of the task (they were using MTM-I analysis). They used the term 'redundancy' to describe the phenomenon.

Globerson (1974) hypothesized that two tasks with the same cycle time, but with different repetition patterns should show different learning rates. He concluded that the learning rate of a task containing repetitions of sub-tasks, is not a constant, but it is positively related to the internal frequency of the elements within the task itself. He also showed that the learning slope of a task approaches the highest slope of any of the elements included in the task and is independent of any of the characteristics of the task. Thus, considering two completely different tasks whose only common feature is that the highest learning slope of *any* element in each task is the same, then, both tasks should approach the same learning slope.

The time for the first cycle, t_1, would appear to be a function of both the overall MTM time as well as the NRT time. It is also influenced by 'job complexity' – but even today, we are not quite sure of the relationship (see Rosenwasser, 1982).

Virtually all experimental research has considered tasks whose overall times are less than 1 minute. But the importance of Human Learning Theory is its application to tasks that are measured in hours and sometimes, days – for example, in missile assembly, CT scanners, aircraft, and so on. Very limited work is reported in this area. Dar-El et al. (1995) have suggested that learning in very long cycle times can be considered as a series of sequential sub-tasks of learning and forgetting, and is therefore treated in more detail in Chapter 5.

2.9. Errors

The facts are, that errors *do* occur during experiments on learning, and are simply ignored. The occurrences of errors are unquestionably reported in work by psychologists, but Buck and Cheng (1993) and Vollichman (1993) are virtually alone in considering errors as an outcome of an emphasis on speed (or, learning) on industrial-type applications.

We often put great pressure on employees to produce high outputs, inviting the potential for increases in errors to occur. This also happened with almost all experimental research. Either the subject is somehow 'penalized' (for instance, an extra time is added to the completion time for that cycle), or else, the occurrence of errors is totally ignored. But no records are maintained to summarize the penalties (see Arzi and Shtub, 1997).

The work of Buck and Cheng (1993) and Vollichman (1993) were done in learning and forgetting, and consequently will be discussed in Chapter 5. But the paucity of experiments that relate learning speeds with quality (through errors), indicates the lack of attention researchers have given to this important area. The sad thing is that the data was there – it was simply not recorded!

2.10. Forgetting

Forgetting occurs most frequently in the industrial setting. For a start, we generally work 8 hours per day, followed by a 16-hour 'break' from our production activities. On weekends, many workers regularly take a 62-hour break. Some forgetting obviously occurs, but the rapid relearning that apparently occurs, does not make these types of breaks a problem for industry. We are much more concerned about longer breaks, e.g., from several weeks to several months, and what it does to production rates. In several industrial studies, it was claimed that a year's break may very well have caused a productivity drop of 60% to 75% (Anderlhore, 1969; McKenna et al., 1985; and Carlson and Rowe, 1976). These are 'serious' numbers that can greatly affect a company's operational costs.

From the viewpoint of industry, forgetting behavior is either ignored, or else, if known to exist, industry has no way of accounting for its occurrence. Early psychological experiments have helped, to some extent, to understand human behavior during the learning-forgetting cycle (see Underwood, 1954, 1968; Wickelgren, 1977, 1981; Christiaansen, 1980; and Carron, 1971). The more important findings were:

- Forgetting increases as the break length increases, and
- Forgetting is lessened as previous experience increases.

Forgetting is a topic fully covered in Chapter 5. We stress its importance because of its prevalence in industrial work, and as a factor to be accounted for in Human Learning theory.

2.11. Continuous Improvement

Work improvement, under the guise of different names, have been with us for over the last half century. Names such as: Work Simplification, Concurrent Engineering, Value Analysis/Value Engineering, Quality Circles, Reengineering and Self-Managed teams are commonplace names given to particular teams whose objectives are always to improve the productivity of organizations. In particular, Quality Circles, and later, the Work Improvement Teams (WIT) and more recently, Self-Managed Work Teams (SMT), have mushroomed in most industries. Their intention is to create an environment for continuous improvement in the organization. Such teams often impact production activities and have the potential for greatly improving the productivity of the manufactured product.

If we're investigating the 'learning effects' of such products, how are we to know if the improvements occur because of improved worker productivity (the acquisition of skill), or, because of induced learning through the work of improvement teams? From the manufacturer's point of view, who cares! The organization aims at improved performances through whatever means. But for the Human Learning analyst, all factors are confounded and the true separation of causes cannot be made.

It is only in the laboratory that the impact of external factors can be controlled since in this environment, there are no external inputs for improvement teams whatsoever.

3 SUMMARY OF LEARNING MODELS • NO FORGETTING

This chapter covers 'no forgetting' learning curve models whose learning curve parameters are determined from field data. However, we must not forget that the ***purpose*** for developing the learning curve is ***'prediction'***. Consequently, it is essential to develop the learning model as early as possible from the start of data generation. This means that potential learning models should focus on evaluating the ***least*** number of parameters compatible with the 'goodness of fit'. As it turns out, our experience shows that 2-parameter learning curve models are about as 'accurate' as models with 3 or more parameters, so consequently this chapter focuses on 2-parameter learning curve models. If one is concerned with fitting a model to large amounts of learning data, then one should not be constrained to working with 'proposed' models – rather, one should apply known mathematical techniques for model fitting to data. Nevertheless, three and more parameter learning models are discussed in Section 3.8.

Learning curve parameter estimation *without* prior experience will be discussed in Chapter 5.

The power model proposed by Wright (1936), although originally applied to completed products, is sufficiently important for many models in this chapter, for it to be presented as if it were originally proposed for individual learning. The same applies to two other learning models – the Cumulative Average Curve and the Stanford B Model. Other review papers on learning curve theory can be found in Engwell (1992), Lane (1986), Hancock and Bayah (1992), Nanda and Adler (1972), Yelle (1979), Belhaoui (1986) and Badiru (1994).

This writer includes in some detail only those models that have 'practical significance', meaning that these should not have more than two learning parameters to evaluate, or, if there is supporting evidence that it has been used in the field (see also Goel and Becknell, 1972). These models are accompanied with fully worked examples. The order in which the learning models are presented bears no significance to their importance.

The models appear in the following order:

3.1. The Power Model

The basic form of the power model is as follows:

$$t_n = t_1 \cdot n^{-b} \tag{3.1a}$$

where,

> n .... is the number of cycles (or repetitions) completed,
> t_n .... is the performance time to complete the nth cycle.
> t_1 .... is the performance time to complete the first cycle,
> b .... is the learning constant.

Taking the logarithm of both sides of equation (3.1) produces an equation to a straight line. Thus,

$$\log (t_n) = \log (t_1) - b \cdot \log (n) \tag{3.2a}$$

Equations (3.1a) and (3.2a) could also be expressed in terms of 'cost', i.e.,

$$C_n = C_1 \cdot n^{-b} \tag{3.1b}$$

and

$$\log (C_n) = \log (C_1) - b \cdot \log (n) \tag{3.2b}$$

where,

> 'n' and 'b' are defined as before,
> C_n ....   is the cost for producing the nth cycle,
> C_1 .....   is the cost for producing the first cycle,

For the rest of the book, we shall only work with the appropriate equations in their 'time' format, though remembering that these may be readily interchanged into their 'cost' equivalent.

Figures 3.1 and 3.2 respectively illustrate the power learning curve and its 'log – log' equivalent. The two learning parameters, 't_1' and 'b', are readily identified and evaluated from Figure 3.2. Thus, 't_1' can be found by its intercept with the ordinate and 'b' is the negative slope of the linear relationship.

A major problem with the power curve is that the performance time approaches zero as the number of cycles becomes large (see Abernathy and Wayne, 1974). Several researchers have proposed models that counteract this tendency - their models will be discussed later in this chapter.

An interesting characteristic of the power curve is that each time production is doubled, the performance time is reduced by the fraction 'b', the learning constant.

Consider two points on the power learning curve., n_1 and n_2. Put $n_2 = 2 \cdot n_1$. Substituting into equation (3.1a) gives,

$$t_{2n1} = t_1 \cdot (2n1)^{-b} \tag{3.3}$$

and

$$t_{n1} = t_1 \cdot (n1)^{-b} \tag{3.4}$$

Dividing (3.3) by (3.4) gives,

$$\frac{t_{2.n1}}{t_{n1}} = (2)^{-b} \tag{3.5}$$

Thus, the ratio in performance times between 'doubled' production is only a function of the learning constant 'b'.

We define , Φ, the *Learning Slope* (or, *Learning Rate*) as,

$$\Phi = 100 \cdot (2)^{-b} \tag{3.6}$$

The value of 'Φ' gives the *rate* of learning. It is the "percent learning" that occurs each time output is doubled. In fitting industrial data to his power model, Wright (1936) empirically found Φ to equal 80%, with its

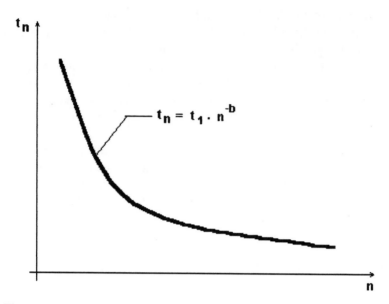

Figure 3.1: The power curve

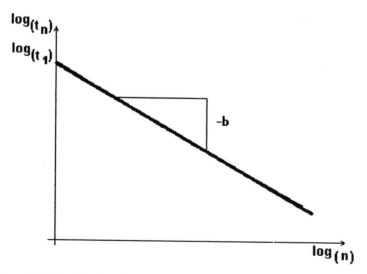

Figure 3.2: The log-log form of the power curve

equivalent 'b' value of 0.322. While there is a tendency for many authors to quote the 80% learning slope almost as a 'mantra', experiments show, depending on 'job complexity', that Φ may vary between 70% and 90% (see Dar-El et al., 1995) and even beyond these limits, from 65% to 95% (see Chapter 4).

In order to predict performance times for the power model, we will need estimates of the two learning parameters, 't_1' and 'b'. With the completion of a few data points of the first few cycles, we need only use the results of two data points in order to estimate the two parameters.

For example, say the first four cycles are completed with performance times as follows: $t_1 = 98$, $t_2 = 83$, $t_3 = 72$ and $t_4 = 69$. Determine the learning curve parameters.

We have six 'pairs' of data which could be checked: '1-2', '1-3', '1-4', '2-3', '2-4' and '3-4'. Let us check pairs '1-3' and '2-4', whose analysis will generate two independent estimates of t_1 and b.

Ordinarily, we would not calculate 'b' and 't_1' in this manner. Instead, we would rely on a much larger data source and apply a regression analysis in order to obtain the parameters. But our illustration has a twofold objective:

a) To demonstrate how 'b' and 't_1' can be calculated.

b) To show we need to make these estimates as soon as possible. We want to use 'b' and 't_1' in a predictive sense – not after all the data is generated.

The same argument will apply for all examples done in this chapter.

Checking '1-3': Applying equation (3.1a) gives –

$$98 = t_1 \cdot (1)^{-b}$$

and

$$72 = t_1 \cdot (3)^{-b}$$

whence,

$$98/72 = (1/3)^{-b}$$

giving

$b = 0.281$.

Checking '2-4': Applying equation (3.1a) gives –

$$83 = t_1 \cdot (2)^{-b}$$

and

$$69 = t_1 \cdot (4)^{-b}$$

whence,

$$83/69 = (2/4)^{-b}$$

giving

$b = 0.267$.

Averaging 'b' gives us, $(1/2).(0.281 + 0.267)$,   i.e., **$b = 0.274$**
The associated learning slope is,

$$\Phi = 100. (2)^{-0.274} ,$$ i.e., **$\Phi = 83\%$**

To find 't_1', we substitute the value of 'b' for all four points and average
the result. Thus,

$$98 = t_1 . (1)^{-0.274}, \quad \text{thus, } t_1 = 98.0$$
$$72 = t_1 . (3)^{-0.274}, \quad \text{thus, } t_1 = 97.3$$
$$83 = t_1 . (2)^{-0.274}, \quad \text{thus, } t_1 = 100.4$$
$$69 = t_1 . (4)^{-0.274}, \quad \text{thus, } t_1 = 100.9. \text{ Averaging, gives } \mathbf{t_1 = 99.2}$$

3.1.1   Finding the total time to complete m cycles, 'T_m'. To calculate
T_m, the total time to complete m cycles, we assume that n is a continuous
function and integrate equation (3.1a); because of this assumption, the
integration is approximate for small n (say, $n \leq 15$), but becomes more exact
for larger values of n (see Camm et al., 1987, for a detailed analysis). Thus,

$$T_m = \int_0^m t_n (n)^{-b} = t_1 \cdot \frac{(n)^{1-b}}{1-b} \Big|_0^m$$

and

$$T_m = \left\{ t_1 \cdot m^{(1-b)} \right\} \cdot (1/1-b). \tag{3.7}$$

Example: Say, $\Phi = 85\%$, $m = 100$ and $t_m = 1.20$ hrs. How long would it take to complete an order for 100 items? We first calculate the value of 'b' from equation (3.6). Thus,

$$85 = 100 \cdot (2)^{-b},$$

whence, **b = 0.234**

From equation (3.1a) we get,

$$1.2 = t_1 \cdot (100)^{-0.234},$$

whence, **t_1 = 3.53 hrs**.

Substituting the values of m, b and t_1 in equation (3.7) gives,

$$T_{100} = \{3.53 \cdot (100)^{(1 - 0.234)}\} \cdot 1/(1-0.234), \text{ giving } T_{100} = \mathbf{156.9 \text{ hrs}}.$$

Working 8 hrs per day, 5 days a week, the operator would take 19.6 days, or, nearly 4 weeks to complete the order.

The question could be asked: How many items would be completed in each of the four weeks?

We answer this question using equation (3.7) with 'm' being the unknown quantity. Let the *cumulative total* number of items completed at the end of the first three weeks be, m_1, m_2, and m_3. We do not need all of the fourth week for completing the order, but $m_4 = 100$.

The time available to complete m_1 items until the end of the first week is, $T_{m_1} = 8.5 = 40$ hrs; for $T_{m_2} = 80$ hrs and $T_{m_3} = 120$ hrs.

For Week 1: Using equation (3.7), gives,

$$40 = \{3.53 . (m_1)^{0.766}\} . (1/0.766), \quad \text{i.e., } \underline{m_1 = 16.8 \text{ items}}.$$

For Weeks (1+2): Using equation (3.7), gives,

$$80 = \{3.53 \cdot (m_2)^{0.766}\} \cdot (1/0.766), \quad \text{i.e., } m_2 = 41.5 \text{ items.}$$

Therefore,   41.5 – 16.8 = 24.7 items were completed in week #2.

For Weeks (1+2+3): Using equation (3.7), gives,

$$120 = \{3.53 \cdot (m_3)^{0.766}\} \cdot (1/0.766), \quad \text{i.e., } m_3 = 70.5 \text{ items.}$$

Therefore,   70.5 – 41.5 = 29 items were completed in week #3.

The remaining 100 – 70.5 = 29.5 items were completed during the fourth week.

3.1.2 To find the average time for completing m cycles. The expression for the Average Time, $\hat{t}_m$ is readily determined from equation (3.7), since we simply divide T_m by m, i.e.,

$$\bar{t}_m = T_m \cdot (1/m) = \{t_1 \cdot m^{(1-b)}\} \cdot \{(1/m) \cdot (1-b)\}$$

or,

$$\bar{t}_m = \{t_1 \cdot (m)^{-b}\}\{1/(1-b)\} \tag{3.8}$$

Thus, for the example just worked out, the average time to complete each item would be,

$$\bar{t}_m = \{3.53 \cdot (100)^{-0.234}\}(1/0.766)$$

i.e., $\bar{t}_m$ = **1.57 hrs.**

3.2. The Cumulative Average Power Model

The cumulative average power model was actually the original form of Wright's proposal when he developed the relationship between the direct labor man hours to the cumulative number of units produced (see also Hancock, 1967).

Instead of t_n used in equation (3.1a), we now use $\hat{t}_n$, the cumulative average time for completing n cycles. Thus,

$$\hat{t}_n = t_1 \cdot (n)^{-b} \tag{3.9}$$

Then, T_n, the total time for completing n cycles is simply going to be 'n . $\hat{t}_n$', i.e.,

$$T_n = n \cdot \hat{t}_n \tag{3.10}$$

or

$$T_n = t_1 \cdot n^{(1-b)}. \tag{3.11}$$

It is slightly more troublesome to find, t_n, the performance time for the nth cycle. Thus,

$$t_n = T_n - T_{n-1}.$$

Finally,

$$t_n = t_1 \left\{ (n)^{(1-b)} - (n-1)^{(1b)} \right\} \tag{3.12}$$

The cumulative average model is often used by researchers when regression values with the power curve are unacceptably low. The cumulative average, by definition, is a continually averaging process that dampens out the effect of 'wild' data points. This characteristic is accentuated as 'n' increases, and as a consequence, important changes in the process can be totally masked by using the cumulative average method. Higher values of R^2 can be expected from the cumulative average method compared to those obtained from using the power model for the same data.

Interpreting data could also be very troublesome. The power of the smoothing process increases as production increases. The data on the most recent production may be totally out-of-character, but the cumulative average curve will not show this (see Conway and Shultz, 1959), i.e., the cumulative average curve tends to dampen out variations to such an extent, that major changes are obscured. Figure 3.3(a) and (b) is a case in point. Figure 3.3(a) is the log-log of the actual data, while Figure 3.3(b) is the log-log form of the cumulative average of the same data. The 'break' in

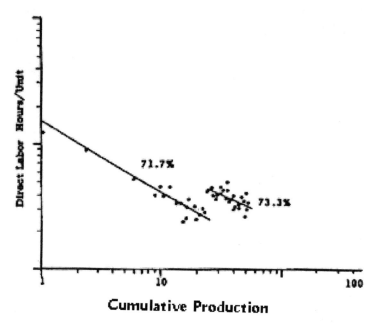

Figure 3.3(a): Final assembly labor, Product D

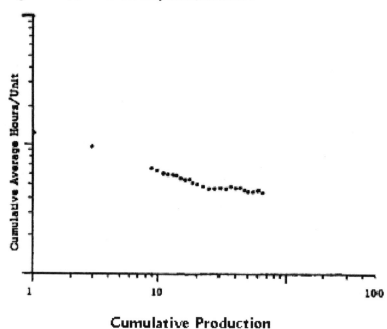

Figure 3.3(b): Cumulative average of final assembly labor, Product D

production was caused by a relocation of the facility to another building, but this effect is completely masked in the cumulative average graph[*].

However, Glover (1965), claims that only by a smoothing process, is one able to get a stable set of parameters for the power curve as a first choice and then move to the cumulative average model in order to improve the R^2 estimates.

The two learning parameters, b and t_1, still need to be evaluated. We do this by applying any two data points to equation (3.11).

Consider the availability of two data points, say, the 4th and 8th cycle.

Let $T_4 = 8.5$ hrs and $T_8 = 12.0$ hrs. How many working hours would it take to complete an order of 150 items?

We use equation (3.11) for evaluating b and t_1. Thus,

$$8.5 = t_1 \cdot 4^{(1-b)}$$

and

$$12.0 = t_1 \cdot 8^{(1-b)}.$$

Dividing these two equations leads to **b = 0.503**   (with Φ = 70.6%). Substituting b in equation (3.11), gives –

$$12.0 = t_1 \cdot 8^{(1-0.503)}.$$

Hence, **t_1 = 4.27 hrs.**

Now we substitute b and t_1 into equation (3.11) to give,

$$T_{150} = 4.27 \, (150)^{(1-0.503)},$$

i.e., **T_{150} = 51.5 hours**.

Working 8 hrs per day and 5 days per week, T_{150} becomes 6.4 days, or, a week and one and a half days to complete the order.

Now we ask: what is the performance time for the 150th cycle (i.e., the last item)?

We use equation (3.12) to obtain the answer. Thus,

[*] Use of copyright material by *J. of Industrial Engineering (IIE Trans)* is gratefully acknowledged.

$$t_{150} = 4.27 \left\{ 150^{0.497} - 149^{0.497} \right\}$$

giving   **t_{150} = 0.17 hrs.**

The learning slope of 70.6% represents a very fast rate of learning, so that the time for the 150th cycle is very low indeed.   In practice, the performance times are likely to be asymptotic to some higher value due to incompressible elements of the task (see Section 3.4).

3.3. The Stanford B Model

A Stanford Research Institute study concluded that the linear formulation of the performance time reduction curves were inappropriate for WWII (World War II) data. Instead, they proposed a modification of the power model as follows:

$$t_n = t_1 \cdot (n + B)^{-b} \tag{3.13}$$

where B is an 'experience factor' which expresses the equivalent units of experience available at the start of a manufacturing program. Generally, B lies between 1 and 10, 4 being a typical value (Garg and Milliman, 1961).

The log-log relationship for the performance time (or, unit labor cost) and n, the cumulative unit number produced, is illustrated in Figure 3.4, where it is seen that for small values of B, the Stanford B model asymptotes fairly rapidly to the regular power model.

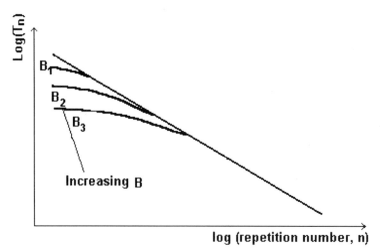

Figure 3.4: Illustrating the Stanford B model

The Stanford B model was successfully applied by the Boeing Company in their aircraft production (see Garg and Milliman, 1961). Studies on 29 cases, showed that 'b' ranged from 0.397 (66.1%) to 0.599 (5.9%) with a mean value of about 0.5. Putting $b = 0.5$, gives,

$$t_n = t_1 \cdot (n + B)^{-0.5}.$$ (3.14)

When $B = 0$, equation (3.14) becomes

$$t_n = t_1 \cdot (n)^{-0.5}$$ (3.15)

which has a 70.7% learning slope.

Because the Stanford B model fails to consider design changes to suit customer specifications (specific application was the Boeing 707), the Boeing Company developed a better fitting modified version of the model by incorporating the number of engineering drawings associated with each product into their calculations (Garg and Milliman, 1961).

Clearly, the Stanford B model was originally intended for describing the learning curve of large products (specifically, aircraft). However, the model was used by Globerson et al. (1989), for describing an experiment on forgetting, where subjects completed 16 repetitions of a task, followed by breaks of varying lengths and then doing another 16 repetitions. For the first experimental run, $B = 0$, while for the second run, $B = 16$. The model was not found to be the most appropriate one to use in that study. Instead, the power model provided improved results.

3.4. DeJong's Learning Model

DeJong (1957) considers the case of tasks containing both manual and machine controlled elements. While the manual parts are compressible with respect to experience, the machine part is not. The production time is divided into two parts – one is subject to the traditional power learning process, while the other remains constant and represents the incompressible part of the task. DeJong defines M to represent the incompressible factor, so that his learning model is expressed as,

$$t_n = t_1 \left\{ M + (1 - M) \cdot n^{-b} \right\}$$ (3.16)

where,

$$M = \frac{\text{Performance time after infinite \# of cycles}}{\text{Performance time for the first cycle}} \qquad (3.17)$$

and,

$$0 \le M \le 1.$$

When there is no machine content, $M = 0$, and equation (3.16) reverts to the power model equation (3.1a) and DeJong's model therefore asymptotes to zero. The misconception of many researchers is in their assumption that DeJong's model considers the *compressible* part of the task asymptotic at some value greater than zero.

There are three parameters to be evaluated in DeJong's model. How should M be determined? Even DeJong isn't much help on this point. He recommends values for m ranging from 0.05 for packaging operations, to 0.75 for metal working operations.

With all cases, he recommends fixing $\Phi = 80\%$ and assumes that 1000 cycles are needed to reach its limiting value (see also Globerson and Crossman, 1976). But both these assumptions are baseless, since we know that b may vary considerably and that learning continues into the millions (Crossman, 1959).

Regrettably, no significant field data is available to support DeJong's model. It is included here since the next model in Section 3.5 redefines the DeJong incompressible factor into a more practical format that enables "M" to be evaluated.

Chapter 4 deals with predicting learning parameters *without* the aid of previous experience. In Section 4.3, a modified definition of "M" can be predicted, and together with predictions for 'b' and 't_1', brings DeJong's model back to life.

3.5. Dar-El's Modification of De-Jong's Model

In Dar-El's modification of the DeJong model, the incompressibility factor is redefined as applying to *all task elements*. Indeed, we can argue that machine controlled activities should not be included in the task to be analyzed; if these occur, then we simply remove the machine times out of the calculations. Thus, equation (3.17) will still apply but with M defined to apply to all elements of the task. So the big question then, is how do we estimate the value of M under the new definition?

We know that PMTS (such as MTM) tables are based on average performance times expected from experienced workers working under "nominal" motivating conditions. For inexperienced workers, these times are only reached after many hundreds of cycles have been completed. We say that the tables are based on "nominal" motivating conditions, since superior operators are expected to perform up to 20% below the MTM standards (see Kargar and Bayah, 1966). The implication is that 80% of the MTM time may be considered as the 'incompressible' content of a task, the amount of which can be determined objectively. Thus, the learning curve would be asymptotic to the "0.8 (MTM)" value, as illustrated in Figure 3.5.

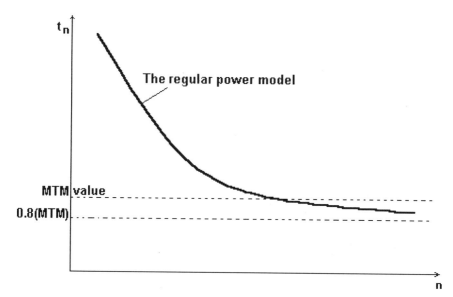

Figure 3.5: Illustrating the Dar-El incompressibility model

Calculations for t_n, T_m and $\bar{t}_m$ are similar to those done for in the regular power curve in Section 3.1.

Put 0.8 (MTM) = A ……. a known value.

By raising the abscissa of the original power curve by 'A', we create a new 'zero' line called '$\hat{0}$' as illustrated in Figure 3.6.
Then,

$$t_n = \hat{t}_n + A \qquad\qquad\qquad (3.18)$$

and

$$\hat{t}_n = \hat{t}_1 \cdot n^{-b}.$$ (3.19)

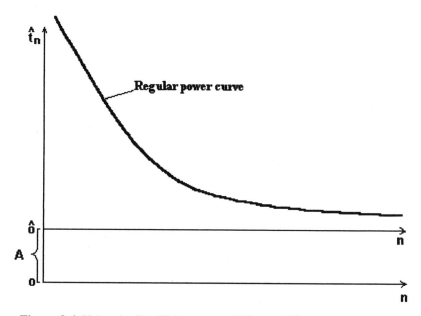

Figure 3.6: Using the Dar-El incompressibility model

From any two data points, the values for b and $\hat{t}_1$ can be found. T_m, the total time for producing m cycles, is given by,

$$T_m = \left\{\hat{t}_1 \cdot m^{(1-b)}\right\}\left\{1/(1-b)\right\} + m \cdot A$$ (3.20)

and $\bar{t}_m$, the average time is given by,

$$\bar{t}_m = (1/m) \cdot T_m = \left\{\hat{t}_1 \cdot m^{-b}\right\}\left\{1/(1-b)\right\} + A.$$ (3.21)

Consider the following example:

The MTM of the task to be performed is 0.75 hrs (use any PMTS system, such as MTM-I, MTM-2, MTM-3, MOST, WORK FACTOR, MODAPS, etc. to determine the task length).
Then, A = 0.8 (0.75) = 0.60 hrs.

Let the job be done four times with the following results:

$$t_1 = 6.65 \text{ hrs} ; \quad t_2 = 5.74 \text{ hrs.} ; \quad t_3 = 5.32 \text{ hrs} ; \quad t_4 = 5.00 \text{ hrs.}$$

From equation (3.15), we calculate the following:

$$\hat{t}_1 = 6.05 \text{ hrs} ; \quad \hat{t}_2 = 5.14 \text{ hrs} ; \quad \hat{t}_3 = 4.72 \text{ hrs} ; \quad \hat{t}_4 = 4.40 \text{ hrs.}$$

As in Section 3.1, we proceed to find b and $\hat{t}_1$ using equation (3.19). Select the following pairs of data points for analysis: '1 & 3' and '2 & 4'.

Pair 1 & 3: $(6.05/4.72) = (1/3)^{-b}$, whence $b_1 = 0.226$,
Pair 2 & 4: $(5.14/4.40) = (2/4)^{-b}$, whence $b_2 = 0.224$.

Hence, Average b = (½) (0.226 + 0.224) = 0.225,   (i.e., Φ is 85.5%)

Check: substitute b for data point 3, in equation (3.19). We get,

$$4.72 = \hat{t}_1 \cdot (3)^{-0.225}$$

whence, $\hat{t}_1 = 6.04$ (we got 6.05 above).
Say, the order is for 125 items. How long will it take to complete the order (working 8 hrs per day and 5 days per week)?
We use equation (3.17) to find T_m. Thus,

$$T_m = \{6.04 \ (125)^{(1-0.225)}\} \ \{1/(1-0.225)\} + 125 \cdot A$$
$$= 328.7 + 75 = \textbf{403.7 hrs.}$$

(i.e., 50.5 days, or, about 5 weeks).
Had we ignored the incompressibility factor, and considered the t_1 to t_4 as is, readers can confirm that the following values are obtained:

$$b = 0.201 ; \quad t_1 = 6.63 ; \quad T_m = 393 \text{ hrs. (c.f., 403.7 above)}$$

i.e., T_m is some 2.3% less than before.

Naturally, the effect of the incompressibility factor is emphasized as the order size increases. For example, had the order been for 2000. Then,

T_m (with incompressibility) = 4850.7 hrs.

and

T_m (without incompressibility) = 4304.5 hrs. (an 11.3% difference).

We have demonstrated that this model has eliminated the main drawback of the power model, where $t_n \Rightarrow 0$, while pertinent calculations remain fairly simple. Of course, when m < 150 cycles, the incompressibility effect is small and can be neglected from calculations.

3.6. The Dar-El/Ayas/Gilad Dual Phase Model

Dar-El et al. (1995), proposed a dual-phase model for learning in individual tasks. Their model was developed when carefully generated research data was found to be poorly matched with all the known mathematically based learning curve models.

Poor fits with the power model was already hinted at when dealing with large numbers of cycles (see Kaloo and Towill, 1979). When the analysis was based on early data, prediction of later cycles tended to be *underestimated* as in Figure 3.7. When the analysis was based on 'late' data points with large numbers of cycles, then the early data points tended to be *underestimated*, as in Figure 3.8 (see also, Cochran, 1960; Yarkoni, 1971; and Rosenwasser, 1982).

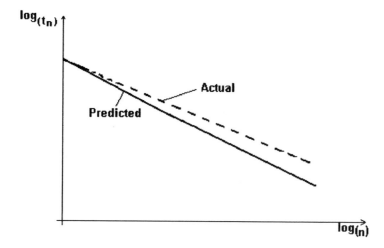

Figure 3.7: Prediction based on early data

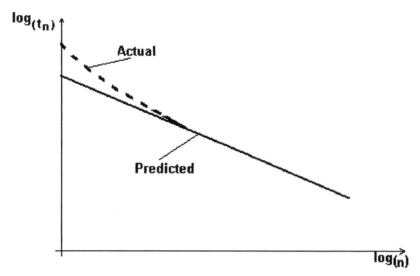

Figure 3.8: Prediction based on late data

Several researchers compensated for these deviations by developing '2-part' learning models – in effect, by altering the value of 'b' at some point in the production as illustrated in Figure 3.9 (see Yarkoni, 1971; Glover, 1965; and Baloff, 1971).

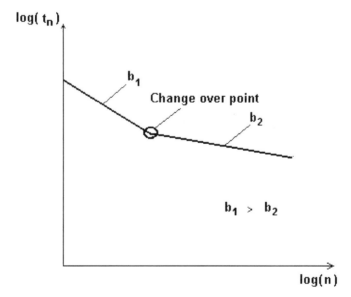

Figure 3.9: A two-linear section learning curve

We argue that these examples of poor fit happen because *two* types of learning occur simultaneously during early cycles. These are 'cognitive' learning and 'motor' learning. We assume that both types follow the power curve – but of course, with vastly different 'b' and 't_1' values.

Cognitive learning includes decision making, following instructions, learning complex sequences, interpreting measurements, searching for cues, etc. Learning tends to be very rapid, i.e., a high value of 'b' and a low value for the learning slope (or, rate), since we expect the worker to rapidly learn what needs to be done. Motor learning on the other hand, is a lot slower (i.e., lower values of 'b' and higher values for the learning slope) and includes all the physical movements needed to accomplish the task.

Initially, performance times are dominated by the cognitive processes. As the cycles increase, there is less influence of the cognitive processes and the motor processes begin to dominate the learning curve, until eventually, only motor learning operates. There are exceptions when some cognitive elements always remain even with large 'n' (e.g., as in pattern recognition) and these should be accounted for (see later). The recognition of cognitive and motor learning was also discussed by Hancock and Foulke (1963), Newall and Rosenbloom (1981), and Fleishman and Rich (1963).

The Dar-El et al. model is illustrated in Figures 3.10a, 3,10b and 3.11, with Figure 3.10a representing cognitive learning, Figure 3.10b – motor learning and Figure 3.11, the combination of the two which results in the dual-phase characteristic of their learning model.

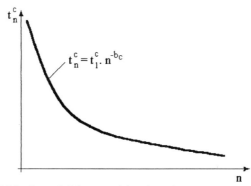

Figure 3.10(a): A model for cognitive learning

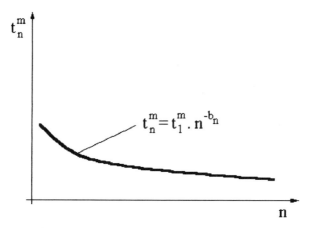

Figure 3.10(b): A model for motor learning

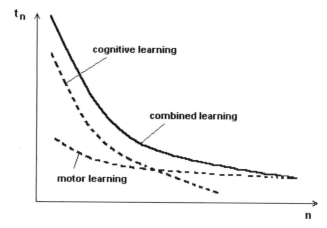

Figure 3.11: Combined cognitive and motor learning

3.6.1 Determining the Learning Parameters. Mathematically, adding two power curves do not result in another power curve (see Conway and Shultz, 1959; and Globerson (1980)) and as a consequence, the 'learning constant' is no longer a constant, but varies with respect to "n". Call this new learning coefficient, b_n^*. The Dar-El et al. model is intended to apply to tasks having high cognitive contents and as with other learning models, it is expected to evaluate learning parameters based on early data points. It is not intended for use when 'n' gets to be too large (n > 500). Besides, this model suffers from the usual drawback with the power curve, in that, the performance times tends to zero for large 'n'.

The new learning model appears to have 4 parameters, "b_c, b_m, ct_1, mt_1", where,

$$b_c \quad \ldots \quad \text{is the learning constant for the pure cognitive processes,}$$
$$b_m \quad \ldots \quad \text{is the learning constant for the pure motor processes,}$$
$$ct_1 \quad \ldots \quad \text{is the first cycle performance time for the cognitive process,}$$
$$mt_1 \quad \ldots \quad \text{is the first cycle performance time for the motor process.}$$

The values of 'b_c' and 'b_m' are constant for all tasks. Based on empirical data (Sabag, 1988), these two parameters have been evaluated as:-

$$b_c = 0.514 \ (\Phi_c = 70\%) \text{ and } b_m = 0.152 \ (\Phi_c = 90\%).$$

Thus, the dual-phase model requires only 2 parameters to be evaluated, namely, 'ct_1' and 'mt_1'.

Describing the combined dual-phase graph gives the following,

$$t_n = (ct_1 + mt_1) \cdot (n)^{-b_m^*} \tag{3.22}$$

where, b_n^* is the instantaneous value of the learning coefficient at repetition 'n'. But also,

$$t_n = (ct_1 \cdot n^{-b_c} + mt_1 \cdot n^{-b_m}). \tag{3.23}$$

Therefore,

$$(ct_1 \cdot n^{-b_c} + mt_1 \cdot n^{-b_m}) = \{ ct_1 + mt_1 \} \cdot (n)^{-b_n^*} \tag{3.24}$$

from which we can express b_n^* as ,

$$b_n^* = b_c - \{1/\log(n)\} \, [\log\{R + (n^{b_c} - b_m/(R+1))\}] \qquad (3.25)$$

where,

$$R = (ct_1 / mt_1). \qquad (3.26)$$

Dividing both sides of equation (3.24) by mt_1, gives

$$R \cdot n^{-b_c} + n^{-b_m} = (R+1) . (n)^{-b_n^*} \qquad (3.27)$$

from which,

$$R = (n^{-b_n^*} - n^{-b_m}).\{1/(n^{-b_c} + n^{-b_n^*})\} \qquad (3.28)$$

From equation (3.27), b_n^* can be expressed as:

$$b_n^* = (1/\log(n)).\{\log (R+1) - \log(R. n^{-b_c} + n^{-b_m})\} \qquad (3.29)$$

In equation (3.29), the value of b_n^* is expressed in terms of R, b_c and b_m. The latter two are constant (and known) and R is constant for a given task, so that b_n^* is a function of 'n' and asymptotes to b_m , the learning constant for motor activities (i.e., $b_m = 0.152$), as illustrated in Figure 3.12.

The fact that b_n^* varies with n has been supported by our experiments (Sabag, 1988), especially during the early cycles. The condition is - that there should be sufficient cognitive elements in the task to influence the value of b_n^* .

In order to evaluate the learning coefficient with respect to 'n', we ensure the availability of early data points, say, up to n = 10 cycles. We assume the learning coefficient is constant over the first 10 cycles and proceed to evaluate, in turn: $b_{n\leq10}^*$, R , ct_1 and mt_1.

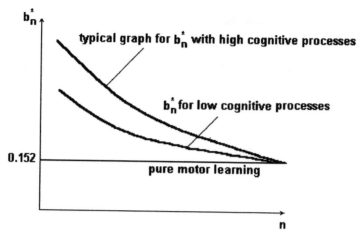

Figure 3.12: Illustrating b_n^* varying with the number of cycles

3.6.2   Calculating the production time for 'm' cycles.   Since the learning coefficient is variable, we simply need to sum up the values of t_n (equation (3.19)), for n = 1, 2, 3,..., m, using a computer spreadsheet, and remembering that the value of b_n^* is obtained from equation (3.29). Thus,

$$T_m = \sum_{n=1}^{m} t_n \qquad\qquad\qquad (3.30)$$

3.6.3 An example. Say two data points ($t_5 = 5.15$ and $t_{10} = 4.02$) are known. Calculate the learning parameters.

To find $b_{n\leq10}^*$: From equation (3.22), we get

$$t_5 = 5.15 = (ct_1 + mt_1) . (5)^{-b_n^*}$$

and $t_{10} = 4.02 = (ct_1 + mt_1) . (10)^{-b_n^*}.$

Then, $(5.15/4.02) = (1/2)^{-b_n^*}$, so that $-b_{n\leq10}^* = 0.3574$ ($\Phi = 78\%$).

To find R: From equation (3.27), check the value of R for n = 5 and n = 10 :-

n = 5:

$$R' = \left(5^{-0.3574} - 5^{-0.152}\right) \cdot \left\{1/\left(5^{-0.514} - 5^{-0.3574}\right)\right\} \quad \therefore R' = 2.00$$

n = 10:

$$R'' = \left(10^{-0.3574} - 10^{-0.152}\right) \cdot \left\{1/\left(10^{-0.514} - 10^{-0.3574}\right)\right\} \quad \therefore R'' = 1.76$$

Averaging   R, we get $R = \frac{1}{2} (2.00 + 1.88)$ **∴ R = 1.88**

To find ct_1 and mt_1 : Dividing both sides of equation (3.22) by mt_1 , we get

$$(t_n/ mt_1) = (R + 1) . n^{-b_n^*},$$

substituting all known values for n = 10, we have

$$(4.02 / mt_1) = (1.88 + 1) . 10^{-0.3574}, \quad \therefore \qquad \mathbf{mt_1 = 3.18}$$

From equation (3.26), $\mathbf{ct_1 = 5.98}$

Check: substitute in equation (3.22) for n = 5,

$$(3.18 + 5.98). 5^{-0.3574} = 5.153 \qquad (c.f. \ t_5 = 5.15).$$

To obtain the learning coefficient for any n >10: For example, what is the expected value of b_n^* , when n = 25 & 250?

For n = 25: Using equation (3.29), we find,

$$b_{25}^* = (1/\log 25).[\log (1.88+1) - \log \{1.88.(25)^{-0.514}+(25)^{-0.152}\}]$$

Therefore, $b_{25}^* = 0.337$ $(\Phi = 79\%)$

This result shows that the learning rate, whose average value for n ≤ 10 was 78%, has increased (become slower) to 79% by the time n reaches 25 cycles.

For n = 250: Using the same procedure, we find $b_{250}^* = 0.302$ ($\Phi =$ 81%). Thus, the learning rate has 'slipped' another 2%.

The example shows that using the Dual-Phase is not much more difficult than in using the power model. The equations appear more complicated but solutions are generated easily using any electronic spreadsheet.

3.7 The Bevis/Towill Learning Model

Bevis, Finnear and Towill (1970) relate the learning curve to an exponential law commonly found in physical systems. They show the output (rather than the performance time) as a function of time. The authors were not the first to use this form. It was proposed several years earlier by Levy, 1965, whose work is discussed in Chapter 8. Using this form for illustrating learning behavior has the advantage of having production asymptote to a maximum level, rather than going to infinity as with the power curve (albeit, as production quantity becomes very large).

The Bevis et al. model expresses the rate of production at time t as R_t, and is determined as follows:

$$R_t = R_c + R_f . \{ 1 - e^{-(t/\tau)} \} \tag{3.31}$$

where,

R_c - is the starting production rate,

R_f - is the steady state production rate during the transient part of the curve,

τ - is the 'time constant', which is the time needed to reach 63% of the steady state production rate.

The Bevis et al. (1970) learning graph is illustrated in Figure 3.13 where relationships between R_t, R_c, and R_f are clearly seen.

Minter (1977) modified the Bevis et al. model by changing 'time' to the number of items produced. Thus, R_n becomes the production rate after 'n' items are produced, and the learning equation becomes,

$$R_n = R_c + R_f . \{ 1 - e^{-(n/N)} \} \tag{3.32}$$

where, N – is the 'quantity' constant.

Three parameters need to be estimated with the Bevis et al. model. R_c must be estimated independently, and the best way is probably to put it equal to $\{1/T_1\}$, where T_1 is found by regression from the power model.

With R_c known, the model requires the minimization of a non-linear equation using production data. The authors use an iterative method for finally determining the parameters (see also Towill and Bevis, 1973).

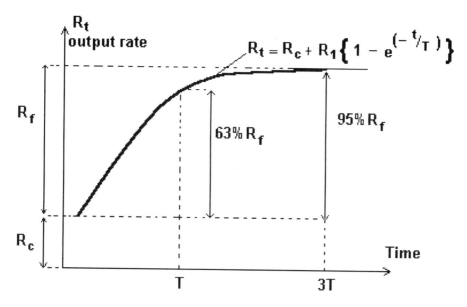

Figure 3.13: Illustrating the Bevis/Towill learning model

Hatching and Towill (1975) developed a similar model that gave estimates for all three parameters, which they claim gave improved results. Cherrington and Towill (1980) found that in order to converge, the linearized models require very good initial estimates of the parameters and a reasonable number of points. They therefore use alternative algorithms which do not require linearization. Later, a similar approach was taken by Donath, Globerson and Zang (1981) for their learning model analysis.

Rosenwasser (1982) compared the Bevis et al. and Minter models with the power models on some 16 sets of data for various learning jobs. Great difficulties were experienced in estimating the three parameters – even using the Hatching and Towill algorithm. Convergence was very sensitive to the initial value assumed for R_c. To obtain better accuracy, R_c was estimated from "$1/T_1$" using the power model. But despite this, it was found that the Hutching and Towill algorithm converged for a wide range of combination values of R_f and τ. Rosenwasser experimented by trial and error to find the best combination to use. But overall, the Power model gave superior results than either the Bevis/Towill or the Minter models.

The Bevis/Towill model was included in this chapter despite the great difficulties experienced in determining the parameters, simply because Towill has developed many more models (discussed in Chapter 8) to fit industrial data. However, because of the difficulty in evaluating the parameters, a worked example is not included in this section.

3.8. Other Learning Models

The learning models appearing in this section (in the opinion of this author) are unsuited to industrial applications. The reasons may be: that there are too many learning parameters to be estimated; the model does not fit the 'real' world; the mathematical 'fit' appears only after all the data points are included in the analysis (what is left to predict?), or else, the model still needs further development.

3.8.1 The Hancock linear model. Hancock (1963) proposed using a linear learning model of the form,

$$t_n = A - B.n \qquad\qquad (3.33)$$

where, A and B are constants.

Hancock never claimed that the linear model should be used in general. It was proposed specifically for very short tasks at levels measured in very few MTM-I elements. Even for very short cycles, the predicted times varied considerably from the actual times for the basic MTM elements (see also Hancock and Foulke, 1963; and Hancock and Sathe, 1989).

3.8.2 The Pegels model. Pegels (1969) proposed a learning model in the form of an exponential function as follows:

$$t_n = \alpha . (a)^{n-1} + \beta \qquad\qquad (3.34)$$

where, α, a and β are parameters determined from fitting the model to all the data.

Buck, Tanchoco and Sweet (1976) suggested a method for estimating the three parameters using methods equivalent to the maximum likelihood estimation method. Later, Buck and Cheng (1993) stated "… it is unclear whether the power form or the exponential series provides a better description of the learning effect".

The parameters are very specific to the fitted data and as a rule, cannot be used in a general sense. However, Buck and Cheng (1993), go into considerable detail in their comparisons between the power and exponential models.

Pegels also presented another model for total labor costs for start-up conditions (see Pegels, 1976).

3.8.3 The Dar-El/Altman learning model. The Dar-El/Altman (1994) learning model was developed an attempt to incorporate the following features:

One) 2 parameters (as in the power model)

Two) allows the learning coefficient, b_n , to be a variable

Three) does not go to zero as $n \Rightarrow \infty$.

In order to achieve this, the learning constant is replaced by a learning coefficient, b_n , which varies with 'n'.

We give b_n the following form:

$$b_n = \alpha - \beta \ln(n) \tag{3.35}$$

This model is still in its development stage, requiring an iterative regression analysis in order to find the optimal value for β.

We compare the R^2 value for both the power curve and the Dar-El/Altman model with data taken from Asseo (1987), Livne and Melamed (1991) and Vollichman (1993), for n = 16 to n = 64. Mean improvements in the R^2 values was in the order of 10%, with the greatest advantage observed for n = 64 (since the power model's tendency to zero). The table below provides comparable results for the n = 64 case. The R^2 value for the proposed model (at 0.957) is far superior to that obtained with the power model (at 0.761).

MODEL	b or α	R^2
Power Curve	0.262	0.761
Dar-El / Altman	0.518	0.957

While showing great promise, the Dar-El/Altman needs further refinement in order to develop simpler steps for deriving its solution.

3.8.4   Yet Other Learning Models. The intention here is to mention other models that have appeared in the literature. All these models require at *least three* learning parameters to be evaluated. As a rule, the authors claim high 'goodness of fit' between their respective model and their data. This can always be expected when fitting a mathematical model to many data points. None of these models are reported in other research, or application.

Cochran (1960) proposed a "S" shaped curve. He claimed that during the early stages of learning, various methods are tried and changes in design are made (including the use of alternative materials). These factors slow down the initial rapid learning rate. As fewer changes enter the system, higher learning rates are expected (see also, Cochran, 1969; Cochran and Sherman, 1982; Carr, 1946).

Carlson (1973) proposed a third degree polynomial in the form,

$$T_n = p + q \cdot n + r \cdot n^2 + s \cdot n^3 \tag{3.36}$$

where p, q, r, and s are empirically derived constants (see also Carlson, 1962).

Glover (1965) introduced a 'work commencement' factor defined as a factor of ignorance at the commencement of a job. However, Glover admits that the factor was seldom used. Glover (1966) also developed a graphical method for controlling the progress of trainees using a log-log plot of the cumulative output v/s the performance time. He claims these graphs, if posted, provide feedback and work as an effective motivator for trainees to maintain their respective learning curve performances.

References to still other work include learning models by: Knecht (1974), Baloff (1971), Buck et al. (1976), Steedman (1970), Bohlen and Barany (1976) and Lippert (1976). Allemang (1977) even proposes to replace the learning curve!

3.9. The Learning Models in Review

It is legitimate to ask which would be the 'best' learning curve model to use in industrial applications. It should be clear that there is no 'best' – the answer is that the model selected should depend on the specific application.

Several comparative studies were carried out. Globerson (1980) compared the power model, the Bevis/Towill (1970) and Linear models. In another study, Rosenwasser compared the power model, the Bevis/Towill model and Minter's adaptation of the Towill model (1977). In both studies the power model came out to be the best. It also had the advantage of requiring only 2 learning parameters to be evaluated (both Towill and Minter require 3 parameters to be determined). Schneider (1989) states that the differences between the power model and the exponential learning models were found to be very small, i.e., the differences in variances were less than 1%.

It is assumed that some early field data is available to evaluate the learning parameters, so that predictions for future conditions can then be made. If applicable, generating more data points should improve the accuracy of the parameter estimates, but remember, the learning constant is averaged over the data range. Also, as far as possible, the outcome of each data point should be recorded separately. However, if one needs to rely on field data, it is very likely the information is only available as cumulative quantities with average times (Conway and Schultz, 1959; Baloff, 1971; Dutton and Thomas, 1984).

As a general guide: if the number of cycles (i.e., the job order) is small (n < 100), the cycle time short (up to 3 minutes MTM time), then use the power curve (3.1), or else, the Cumulative Average model (3.2). For larger orders (n > 100) choose the Dar-El Incompressibility model (3.5), since we will also need to determine the MTM time for the task. Alternatively, if the mathematical manipulations are not too daunting, the Bevis/Towill (1970), or Minter (1977) models may be used, though remember, we need to evaluate 3 parameters with these. If the data appears to be erratic, or, if the R^2 is too low (< 75%), consider using the cumulative average model.

If the task contains high cognitive elements, but orders are low (and the learning coefficient is suspected to be variable), use the Dar-El et al. Dual-Phase model (Section 3.5). For large orders with high cognitive elements, choose the Dar-El incompressibility model or else, the cumulative average model (this should nicely mask the effects of changing 'b').

All of the learning models mentioned above (except the Bevis/Towill and Minter models) have been presented with worked examples, and therefore should present no difficulties in their application.

The real problem in learning curve application is the ability to predict the learning parameters at the planning stage when **no** field data points are available, i.e., with no previous experience. This aspect will be discussed in the next chapter.

The next chapter will also show how the DeJong model can be resurrected, which could then become an attractive alternative for many industrial applications.

4 DETERMINING THE POWER CURVE LEARNING CURVE PARAMETERS

The two power model learning parameters, 't_1' – the performance time for the first repetition, and 'b' – the learning constant, were discussed in Chapters 1 and 2.

We wrote of the difficulty in obtaining estimates of 't_1'; because of its high variance, the performance time for the first cycle must be considered an unreliable estimator of 't_1'. The examples given in Chapter 3 utilized early data in order to demonstrate the calculation methods with the various learning models. It is quite likely that using only a few data points (e.g., up to 6 repetitions), are likely to give unreliable estimates of 't_1' and 'b'.

In practice, if we were relying on past data for obtaining learning parameter estimates, then using the "log-log" graph would most likely give the best results, since in one step we can find both 't_1' and 'b'.

Finding a value for 'b', the learning constant, appears to be a lot easier than estimating, 't_1', the time for the first cycle. There are now several tables published on 'b' values found for different kinds of work (mechanical assembly, electronic assembly, machining operations, packaging goods, etc.), so one may be tempted to select a value based on a "best fit" description between the job on hand and an appropriate table description. Tables 4.1, 4.2 and 4.3 are taken from various sources in which learning slope 'b' values are quoted for specific types of work.

Moreover, a survey of learning slopes by Baloff (1971) shows these varying across industries, processes and products – even for the same product in the same plant (Nadler and Smith, 1963).

Table 4.1 Learning slopes from reported studies

Application	Learning Slope %
Truck body assembly (Glover, 1965, 1966)	68
Electronic lab. assembly – the sine curve (Sabag, 1988)	70
Electronic assembly – fine locations (Livne and Melamed, 1991)	70
Process control simulation – no quality specification (Vollichman,1994)	73
Complex electronic assembly 'product A' (Conway and Schultz, 1959)	73
Electro-mechanical assembly (Brenner, 1995)	73 - 77
Paper – days/10,000 tons (Morooka and Nakai, 1971)	74 - 75
Assembly – 'Erector Set' toy (Bailey, 1989)	74
Electro-mechanical assembly 'product B' (Conway and Schultz, 1959)	75
Apparel –various (Baloff, 1971)	75 - 79
Total hours/small ship (53 ships) (Morooka and Nakai, 1971)	76
Process control simulation – medium quality (Vollichman, 1994)	77
Apparel manufacture (10 products) (Rosenvasser, 1982)	77 - 85
Service time IBM electronic machines (Kneip, 1965)	79
Computer data entry (Shtub et al., 1993)	79
Keyboard entry on business machines (Kilbridge, 1959)	80
Identifying shapes on computer screen (Arzi and Shtub, 1997)	80
Automobile assembly – manual (Baloff, 1971)	80 - 84
Service calls /150 IBM machines (Kneip, 1965).	81
Electro-Mechanical assembly (Morooka and Nakai, 1971)	82 - 89
Assembly – small electrical switch (Livneh and Melamed, 1991)	82
Servicing automatic transfer machines (Glover, 1965, 1966)	83
Electro-Mechanical assembly inspection (Morooka and Nakai, 1971)	83 -89
Cigar making (Crossman, 1959)	84
Process control simulation – high quality	84

(Vollichman,1993)	
Final assembly on 'product B' (Conway and Schultz, 1959)	84
Total hrs/boiler, tank and pipes (small ships) (Morooka and Nakai, 1971)	84 - 88
Labor hrs/ gun barrel on boring machine (Andress, 1954)	85
Mechanical assembly (3 studies) (Brenner, 1995)	84 - 88
Service time- IBM electro/mechanical machines (Kneip, 1965)	87
Cost of purchased aircraft subassemblies (Wright, 1936)	88
Musical instruments – (Baloff, 1971)	88 - 91
Automobile assembly – jigs (Baloff, 1971)	89 - 90
Computer tracking (Arzi and Shtub, 1997)	90
Electronic assembly – location unimportant (Sabag, 1988)	90
Raw Material usage/aircraft (Wright, 1936)	95

Table 4.2: Learning slopes for machining operations

(based on Nadler and Smith (1963)[*] and others (refs given))

Application	Learning Slope (%)
Operations in a large steel fabricating plant:	
(also Titleman, 1957) Gas cutting – thin plates	85
Gas cutting – thick plates (Machine)	92
(also Titleman, 1957) Shearing plates	82
Grinding – manual	98.5
(Morooka and Nakai, 1971) Grinding for castings	94 - 96
Milling	98
(Morooka and Nakai, 1971) Lathe	84 - 87
Assembly	83
(Morooka and Nakai, 1971) Automobile Assembly	83 - 84
Fitting	83
Welding – manual	88
(McCambell and McQueen, 1956) Burring, Sanding	82
(Konz, 1995) Drill, Ream and Tap	87
Drilling	87
(Barnes and Amrine, 1942) Screwdriver work	95
(Bapu, 1967) Punch press	95
(Morooka and Nakai, 1971) Idle/Loss time	76 - 83
Shop groups:	
Sheet metal and machine shop	76 – 87

(Morooka and Nakai, 1971) Small ships: machining	86
Primary assembly and sub-assembly	70 – 83
(Morooka and Nakai, 1971) Ass'y – cash registers	92 - 96
Data from one firm:	
Electronic assembly – final	72.9
Electro-mechanical assembly - final	75.9

*Use of copyright material by the *Int. J. Production Research* is gratefully acknowledged.

Table 4.3: Learning slopes for various activities

(based on material used by the Israel Defense Industries–Systems Eng'g.)

Industrial Process	Learning Slope %
Assembly of complex machine tools – new models	75 - 85
Repetitive electronics manufacturing	90 - 95
Repetitive machining/punch press operations	90 - 95
Repetitive clerical operations	75 – 85
Repetitive welding operations	90
Purchased parts	85 – 88
General assembly	80 – 85
Operators, using standard methods on repetitive work	85
Setups, repair of rejects, nonrecurring activities	80
Subsystem	Learning Slope %
Fabrication	85 – 95
Subassemblies	75 - 85
Final assembly	75 – 80
Structures subsystem	84 – 88
Electronics subassembly	85
New state-of-the-art product	75 - 80

Determining a value for 't₁' is another matter altogether. There are simply no methods available! Using MTM (Work Factor, MODAPS, etc.) values yield numbers that are simply too tight, since they're to be applied for *experienced* workers. Using Globerson's NRT definition, gives even tighter results. The best researchers have done in the past was to estimate (guess, more likely!) the number of cycles needed to produce manufacturing rates equivalent to that done by experienced workers. But, even here, we must rely on pure guesswork, since both Globerson and Crossman, and DeJong proposed that 1000 cycles was the magic number, indicating that Standard Times were achieved after completing 1000 repetitions of *any* task! This assumption enabled the researchers to produce some results – but practitioners, in relying on common sense, have simply ignored this approach. Hence, the reliance in practice on doing the actual job in order to generate some data which could then be analyzed.

However, if one can use *real* data for finding the parameters, then using the tables are simply not justified, since the estimates based on actual data are going to be far more reliable.

If the parameters are to be used for prediction purposes (**without** simulation), then better methods are available which are discussed in this chapter.

Parameter determination will be discussed as follows:

    4.1   Determining parameters from actual data
    4.2   Predicting learning parameters *without* prior experience
        4.2.1   The learning constant 'b'
        4.2.2   Performance time for the first cycle 't_1'
    4.3   Reviving DeJong's Learning Curve Model
    4.4   Predicting the number of cycles to reach standard
    4.5   Assessing "Previous Experience".

4.1. Determining Parameters from Actual Data

This is the simplest situation, since all we do is to analyze the data on a "log-log" graph to immediately obtain estimations of both parameters. The reader will understand, when learning models require more than two parameters, the analytical process becomes more complex and the question as to how many data points are needed, must be addressed. This is the main reason for keeping away from models that require more than two parameters to be estimated (see Camm, 1985).

Analyzing pairs of data points (as was done in the examples for illustrating the various learning models in Chapter 3) may not be the most efficient way for obtaining the two parameters, since we must also have statistical validation on the number of independent 'pairs' needed to obtain acceptable estimates.

The reader is cautioned against including steady-state data in the log-log analysis as it may easily distort the value of 'b' (see Conway and Schultz, 1959; and Baloff, 1971).

4.2. Predicting Parameter Values *Without* Prior Experience

The main value of learning curve models is being able to use these for predictive purposes at the design, cost estimation and planning stages. By implication, we will have no 'previous experience' to fall upon in order to find the two learning curve parameters. How to proceed will now be discussed.

4.2.1 Predicting the learning constant "b"[*].

As mentioned earlier, all practitioners could do was to find a 'near' match between their task and one appearing in one of the published tables. What if a good match could not be found? What of the reliability of the data found in these tables? So, one simply guesses a value and proceeds! This appeared to be the formal approach for finding 'b' without prior experience.

This author has supervised several graduate students charged with the task of developing new ways for determining the value of 'b' (as well as 't_i'). Rosenwasser (1982), working with Job Complexity, appeared to have the most promising approach using the Information Content of the task, but his research needed a little more to meet the needs for good scientific research and his data remains incomplete to this day!

However, excellent results were obtained via Sabag (1988) and published in Dar-El et al. (1995). The approach taken was to reduce the task into its constituent smaller sub-tasks, and then to roughly estimate its 'b' value by analyzing its job characteristics.

Recall in the Dual-Phase model (Chapter 3, Section 3.6), it was mentioned that empirically determined 'b' values for 'pure' cognitive and 'pure' motor activities were given as "0.515 ($\phi = 70\%$)" and "0.152 ($\phi = 90\%$)" respectively (see Dar-El et al., 1995). Experimental work by other researchers (see Tables 4.1 to 4.3) have obtained 'b' values that suggest the range could be extended, say, by up to 5% in either direction, i.e., ranging from 65% to 95%. But back to Sabag; the 70% to 90% range, assumed to cover the learning slopes of all activities, is divided into four sections as shown in Figure 4.1. Our task then, is to allocate each subtask to its most appropriate sector which is designated as follows:

C2 - "Highly Cognitive" - average ϕ is 72 ½ % ,
C1 - "More Cognitive than Motor" - average ϕ is 77 ½ % ,
M1 - "More Motor than Cognitive" - average ϕ is 82 ½ % ,
M2 - "Highly Motor"   - average ϕ is 87 ½ % .

---

[*]Use of copyright material by *IIE Transactions* is gratefully acknowledged.

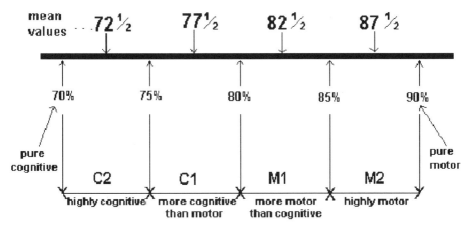

Figure 4.1: Scale of learning slopes and the four classifications

It is a lot easier to classify the 'b' value of a subtask into one of these four categories then trying to accurately determine its value. If a subtask is shown to have a 'b' value of 81.6%, then this task will be classified as belonging to M1 with a slope of 82 ½ %. Such 'errors' can be as high as 2½%, but we argue that in practice, these errors would tend to balance out since we assume each overall task is made up of many subtasks whose combined learning slopes determine the learning slope for the overall task.

The subtasks are classified into one of the four categories according to the skills required for executing the work. This is achieved through the use of a set of six questions, each of which require a 'Yes' or 'No' answer. The questions are based on characteristics associated with cognitive processes, as well as factors affecting the level of manipulative difficulty likely to be encountered.

The six questions are as follows:

Q1. Does the operator need to refer to documentation (manuals, drawing instructions, etc.) in order to perform the operation even after many (say, 50) repetitions are made?

(*an addition*) For those tasks not requiring physical assembly: Does the operator require rapid tracking, or motoring, with special counteracting maneuvers for situations that occur quite frequently?

Q2. Does the operator need to confirm the correctness of his performance at several stages of the task, one or more of which may even require the determination of a condition in order to respond accordingly (e.g., to make an adjustment)?

Q3. Does the task require a high level of manipulative skills for its performance?

Q4. Does the task require a medium level of manipulative skills for its performance?

Q5. Is a mandatory inspection, or review, required on completion?

Q6. Could the operator readily memorize the sequence of activities within five repetitions?

The questions were developed by defining those characteristics that represent cognitive activities (sequencing, methods, measurements, decision-making, etc.), as well as those requiring varying levels of manipulative dexterity. The nonexistence of these characteristics would therefore mean that the task is essentially a 'motor' one.

The questions are ranked first by activities associated with higher cognitive characteristics, in the following manner:

Q1 is particularly suited to electronic assembly where locations and sequence of activities are complex and take many cycles to learn. Here, the operator is likely to constantly refer to documents, drawings, specifications, etc., in order to accomplish the task, even after 50 products are completed. Additional characteristics include tracking or monitoring of data or targets, which would require some counteractions to be used to keep the situation under control. This category would be ideal for combat pilots, navigators, counteracting missiles that are launched, and so on.

Q2 meets the requirements of electronic work, fine mechanical work, etc. where progress inspections, or reviews are needed that may necessitate adjustments. A combination of positive answers to Q1 and Q2 would be sufficient to categorize the task as C2 or, "highly cognitive".

Q3 asks if the task requires a high level of manipulative skills such as found in very accurate, fine locations for assembly, or else, requiring the manipulation and orientation of very fine components. These activities require both high cognitive and motor skills in their execution. Combinations of 'Yes's' with Q3, with either Q1 or Q2, are classed as C2. A 'Yes' for Q3, without Q1 or Q2, is given the C1 category ("more cognitive than motor"), since the cognitive elements in the Q3 type task is assumed to be able to be learnt within a reasonable number of repetitions. The C1 category is also given to a task having a 'Yes' for Q2 and a 'No' for Q1 and Q3. All other combinations of answers require answers to further questions.

Q4 considers whether the task requires a medium level of manipulative skills. This level would, for example, suit a task of knotting threads, as required in the textile industry, assembling small electro-mechanical components into snug locations, etc. A 'Yes' to Q1 and Q4 is given a C1, whereas a 'Yes' to Q1 and a 'No' to Q4 requires further questions to be answered.

Q5 requires a mandatory inspection on completion of the task. This can take the form of a measurement for conformance (size, tolerance, weight, concentration, etc.), or else, to ensure that parts fit correctly in order that the next activity can be started. A 'Yes' to Q1 and Q5 is given a C1 category, while a 'Yes' to Q1 but 'No' to all others (up to Q5) is given a M1; because once the sequence of events is learned, assembly can proceed without difficulty.

The last question, Q6, checks whether an operator can memorize the sequence within a short number of repetitions (say, five). Note that Q6 is a much milder form of Q1, but answering 'No' still leaves the task in category M1.

Tables 4.4 and 4.5 summarize the category types associated with the combination of answers to the six questions.

Table 4.4: Categories according to answers to Q1, Q2 and Q3

Question #	Answers	to	Questions				
Q1	Y	Y	N	N	N	Y	N
Q2	Y	N	Y	N	Y	N	N
Q3		Y	Y	Y	N	N	N
Decision	C2	C2	C2			Ask Question Q4	
				C1	C1		

Table 4.5: Categories according to answers to Q1 to Q6

Question	Answers to Questions		Question	Answers to Questions	
From Table 4.1 Q1	Y		Q1	N	
Q2	N		Q2	N	
Q3	N		Q3	N	
Q4	Y	N	Q4	Y	N
Decision:	C1	Ask Q5	Decision:	M1	Ask Q5
Q5	Y	N	Q5	Y	N
Decision:	C1	M1	Decision:	M1	Ask Q6
Q6			Q6	Y	N
			Decision:	M2	M1

The validity of the questionnaire was checked against several tasks involved in various experiments done by Sabag (1988), Brenner (1995), and Livne and Melamed (1991). These are reported in Table 4.6, whose results indicate that the selected categories are totally consistent with respect to the actual learning slopes that were obtained in the experiments. Some residual rounding-off errors do exist, but it will be shown that these errors tend to balance out for long cycle time tasks composed of a large number of short cycle time tasks.

A sensitivity analysis of errors made in the learning slopes classification process (Sabag, 1988) supports our contention that with long cycle time jobs, the averaging process sharply reduces the overall error. A theoretical task comprising 60 subtasks was subjected to classification errors, at levels given in Table 4.7. It is observed that the resultant errors in the learning slope estimates is a maximum of 6.5% for the worst case scenario – where *all* errors are in the same direction.

Table 4.6: Comparing learning slope categories with measured values

Source		Predicted Category	Predicted Learning Slope	Exp'tal Learning Slope	Difference Actual – Predicted)
Sabag	(i)	C2	72.5	74.6	+ 2.1
	(ii)	M2	87.5	87.9	+ 0.4
Brenner	(i)	M2	87.5	87.7	+ 0.2
	(ii)	M1	82.5	84.7	+ 2.2
	(iii)	M1	82.5	83.5	+ 1.0
	(iv)	C1	77.5	75.3	- 1.8
	(v)	C1	77.5	77.4	- 0.1
	(vi)	C2	72.5	73.2	+ 0.7
Livne	(i)	M1	82.5	81.5	- 1.0
and	(ii)	M2	87.5	90.3	+ 2.8
Melamed	(iii)	C2	72.5	70.0	- 2.5

Table 4.7: Sensitivity errors in the classification process

% Error in Classification	% Error in the Total Time
± 20	2.6
+ 20	2.2
- 20	6.5

There are no problems in extending the learning slope scale from 65% to 95% and having another two categories: "ultra pure" cognitive (C3) whose mean value is 67.5%, and "ultra pure" motor (M3) whose mean value is 92.5%. Doing this in no way detracts from the steps taken thus far for predicting the slope 'b'.

4.2.2 Predicting the performance time for the first cycle. It has been a far more difficult task to predict a value for the first cycle, than for predicting the learning constant, 'b'. What could be a basis for its prediction? Clearly, 't_1' is a function of both the task time and job complexity, i.e.,

$$t_1 = f \text{ (task time, job complexity)} \qquad (4.1)$$

but how to determine the relationship? Once again, the work by Rosenwasser (1982) based on the Information Content appeared to be a promising approach, but as mentioned earlier, the experiments were never satisfactorily completed.

Previously, the 'best' approach was to determine 't_1' indirectly, by guessing how many repetitions were needed to achieve the Time Standard (e.g., the MTM time).

Let 'G' represent the number of cycles needed to achieve the Time Standard. Then,

$$t_G = t_1 \cdot (G)^b \qquad (4.2)$$

If 't_G' were equal to the MTM standard (let's call this value 'C'), then by guessing the value of G and picking up the 'best' fit for 'b' from Tables 4.1 to 4.3, one could proceed to find an estimate for 't_1'. Thus,

$$t_G = C = t_1 \cdot (G)^b \qquad (4.3)$$

So what value of G should we take? Both DeJong (1957) and Globerson and Crossman (1976) assume a value of G = 1000 cycles; but there is no basis for their assumption, since G cannot be a constant.

It was left to Sabag (1988) and reported in Dar-El et al. (1995) to come up with a solution. This was based on the dimensionless number (t_1/C), which we knew was influenced by Φ, the learning slope. Accordingly, (t_1/C) was plotted against Φ for all available (and reliable) research data (taken from Hancock and Foulke, 1963; Rabinovitch, 1983; Sabag, 1988; and Brenner, 1990). What we looked for was reliable research data for tasks whose MTM standard times were given and whose 'b' and 't_1' estimates were obtained from experiments.

The resulting plot in Figure 4.2 shows a fairly strong linear relationship between (t_1/C) and Φ, the learning slope. Figure 4.2 contains some 22 data points with a linear regression, R = 0.96 which is sufficiently good for obtaining a first estimate of (t_1/C).

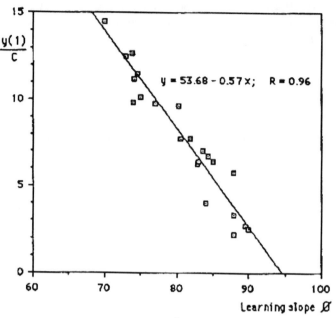

Figure 4.2: Relationship between t_1/C and Ø (from Dar-El et al., 1995)

Having predicted the value of 'b' (or, Φ) as described earlier, one merely needs to read the appropriate value of (t_1/C) from Figure 4.2 in order to find 't_1', since by multiplying (t_1/C) by C, gives us 't_1', i.e.,

$$t_1 = (t_1 / C) \cdot C. \hspace{3cm} (4.4)$$

The value of C is known (from MTM, MOST, WORK FACTOR, MODAPTS, etc.) and the value of 't_1' is found. Comparisons between predictions for 't_1' using this approach and actual 't_1' estimates for the tasks mentioned in Table 4.7 are given in Table 4.8 below.

The estimate of 't_1' for the M2 category is the poorest, suggesting that perhaps the location of 'pure motor learning' should have a higher slope than assumed. Ignoring the M2 data, leaves an average difference of only 4.9% for the C2, C1, and M1 categories, which is a remarkable level of accuracy. However, even taking the M2 data into account, the 14% average absolute difference between experimental and predicted values is still fairly good for planning purposes.

Table 4.8: Analysis of the accuracy for predicting 't_1'

Source		Predicted Category	Φ	(t_1/C)	Actual C	Actual 't_1'	Predict 't_1'	Diff. %
Sabag	(i)	C2	72.5	12.6	151.8	1774	1913	-7.8
	(ii)	M2	87.5	4.1	88.8	288	364	-26.4
Brenner	(i)	M2	87.5	4.1	23.9	65	98	-50.8
	(ii)	M1	82.5	6.9	25.0	160	172	- 7.5
	(iii)	M1	82.5	6.9	21.5	128	148	-15.6
	(iv)	C1	77.5	9.7	36.0	363	239	+ 3.9
	(v)	C1	77.5	9.7	15.0	146	146	0
	(vi)	C2	72.5	12.6	15.0	187	189	- 1.1
Livne	(i)	M1	82.5	6.9	15.0	91	104	-14.3
and	(ii)	M2	87.5	4.1	34.8	116	143	-23.3
Melamed	(iii)	C2	72.5	12.6	18.0	237	227	+ 4.2

The strong linear relationship seen in Figure 4.2 is both logical and powerful in its implications. The time for the first repetition in a highly cognitive task is likely to take about 13–15 times longer than its standard time, whereas the first cycle time for a highly motor task would take about 2½ times as long as its standard time. Clearly, there is much more to improve with highly cognitive tasks than with highly motor tasks.

These results may also explain why short run assembly lines for new R&D products invariably experience problems with respect to completion time.

4.2.3 Reviewing the methodology for predicting 'b' and 't_1' values.
Knowing 'b' and 't_1' provides all the information needed for predicting the learning characteristics of industrial tasks *before* the commencement of production. To recapitulate, the method used for predicting learning curve parameters is as follows:

(One)   Define the task or activity. Use the questionnaire to determine its representative category and 'b' value.

(Two)   With 'b' known, find the appropriate value of (t_1/C) from Figure 4.2.

(Three) Estimate the Standard time, C – use any PMTS system.

(Four) Multiplying (t_1/C) by C will give the value of 't_1'.

Φ (or, 'b') and 't_1' are now known.

Keep in mind that in Dar-El et al.'s (1995) Dual-Phase model, 'b' becomes a variable, gradually altering it's value until it asymptotes to $b_m = 0.154$ for very large n. Thus, the method developed for predicting the learning curve parameters are not applicable for n large. But in most present day applications, we deal with production quantities in the 100's and not in the

100's of thousands. Therefore, the methods developed remain valid, and can be used with several learning models given in Chapter 3, including: the Power Model, the Cumulative Average Model, Dar-El's Modification of DeJong, and the Stanford B model (if M were known). It will also bring to life, DeJong's model, as will be discussed in the next section.

4.3. Resuscitating DeJong's Learning Curve Model

Referring back to Chapter 3, Section 3.4, it was noted that DeJong's learning model was described as appealing in that an incompressibility factor, M, was included in his equation. But this required the DeJong model to evaluate three parameters: " b, t_1, and M". His original definition of M applied to the 'machine time' included in the overall task (in the 1950's most learning curve studies dealt with mechanical or, electrical processes which included machine time), and even DeJong was unable to evaluate M. Consequently, any review of learning curve literature, would refer to DeJong's model as 'appealing' – but in reality, it was *useless* – because of our inability to find an accurate value for M.

Today, some 45 years later, learning curve research and industrial applications do not even mention "machine hours" being an element in their studies. Thus, if we were to redefine DeJong's incompressibility factor (M) as applying to the whole task (and ignoring 'machine time'), then the problem is solved!

It is solved because M is the *inverse* of the ordinate in Figure 4.2, i.e.,

$$M = C/t_1 \tag{4.5}$$

and ' t_1/C ' is obtained directly from Figure 4.2. Thus, for highly cognitive work, M would have a value of about 0.07, whereas for high motor activities, M would be around 0.4 and there is no need for any guesswork!

The parameter values for DeJong's model: "b, t_1 and M", are now all predictable, and DeJong's model can be added to the list of 'practical models' for individual learning. Indeed, because of its 'appeal', DeJong's model may become the popular method to use for large orders, since performance times never approach zero). DeJong's model can also safely replace 'Dar-El's modification of DeJong's model' (Section 3.4) as a preferred model to use.

4.4. Predicting the Number of Repetitions to Reach Standard

Determining the value of 'G', the number of repetitions to reach the Standard Time (e.g., such as the MTM time) is a simple matter once 'b' and 't_1' are known. We merely substitute appropriate values for the terms in equation (4.3), where $C = t_1 . (G)^{-b}$. Thus,

$$(G)^b = (t_1 / C).$$　　　　　　　　　　　　　　　　(4.6)

Table 4.9 shows the G values for the average of the six categories (we are now adding the two 'extremes') 'C1 to C3' and 'M1 to M3'. The values of 'G' in Table 4.9 apply to cycle times that are no more than about 10 minutes, otherwise, forgetting begins to play an important role in the learning process, and the 'G' values in Table 4.9 will not apply. The learning process with long cycle times is treated in Chapter 6, Section 6.3.

Table 4.9: Number of cycles to reach Standard

Learning Slope Category	Φ (%)	t_1 / C	'G'
C3	67.5	15.2	85
C2	72.5	12.4	227
C1	77.5	9.5	454
M1	82.5	6.7	936
M2	87.5	3.8	1009
M3	92.5	1.0	indeterminate

4.5. Assessing Previous Experience

Assessing 'previous experience' is an exceedingly difficult task, which explains why *not one* relevant research paper exists on this topic!

How does one begin to assess 'previous experience' – especially when the new task has never been done before, but has some similarity to tasks done in the past?

Some assessment procedure is used whenever we select personnel for our plants. What do we look for in the selection process? Is it only aptitude, motivation and training, or, is 'previous experience' important? If it is, how do we assess this?

Each new job contracted to be done by the plant, is basically comprised of two kind of activities: work that is very new to the plant and work that bears similarity (even identical) to other jobs done previously in the plant.

New work is easy to deal with, since it fits our learning model concept which assumes the use of "naïve" operators. We should have no problem in determining the parts of the new job that are actually new. Then, its analysis should be dealt with by predicting the learning parameters (see this chapter) and then to select an appropriate learning curve model to describe the learning experience.

It's the second type (where similar activities have been done) which needs to be assessed *and* which turns out to be the most prevalent situation found in industry.

Consider what factors may influence the decision for assessing 'previous experience'. One factor would surely be the "degree of similarity" that may exist between the new work and our previous experience. Another obvious one would be "the number of years experience with similar work". There are certainly other psychological factors, such as aptitude and motivation, which play a role, but dealing with the two factors above appear to be a good starting point.

At this stage, the only approach one can take would be a heuristic one, which tries to capture the two factors in some meaningful way. It is proposed to describe the "degree of similarity" through four broad categories as given in Table 4.11.

Table 4.11: Four categories for describing the "Degree of Similarity"

Category 1	95% or more, identical work.	No training or instructions needed. Physical and Mental processes virtually identical. Transfer of skills 100%
Category 2	75% classed as similar work.	About ¾ of operator's work classed as being "very" similar. About ¼ of the work has 'some' similarity.
Category 3	50% classed as similar work.	About half of operator's work classed as being "very" similar. The rest of the work bears 'some' similarity.
Category 4	25% classes as similar work.	About 1/4 of operator's work classed as being "very" similar. 75% of his work bears 'some' similarity.

Next, we consider the heuristic part: What equivalent previous experience should be given for spending a given number of years in a specific category? We will consider each category in turn.

4.5.1 Category 1. As we mentioned earlier, we should have no problems with this type of work since we have a full 100% transfer of skills.

Depending on the operators years of experience in the particular task, we can 'pick up' his position on the learning curve and transfer this point to be the starting position on the new work as illustrated in Figure 4.3. It is observed that the starting 'experience' remains the same as his performance times when he left off from the previous job.

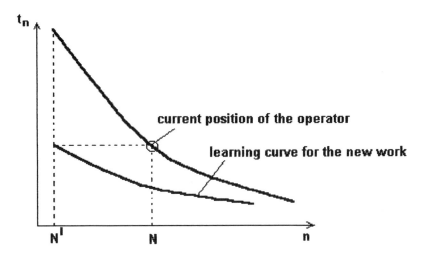

Figure 4.3: Accounting for previous experience

In effect, the learning curve simply continues with the starting point at N' which must be approximately estimated from production records (one can err on the side of the operator by underestimating this number, without having to pay too high a price).

Example: A new job is to be started – it is identical to the work being performed today (except for very minor inconsequential changes), and one of our regular operators will be put on this new job.

From factory records, we make an assessment of this operator's learning slope 'b', and the performance times he was turning in on his recent work. We 'roughly' estimate that the operator has done about 3,500 items over the five years experience he has. We reduce this experience to 90% because of the small changes in design and methods.

Starting point for the operator is therefore fixed at 90% . 3,500 = 3,150, and we further round it down to 3000 items. Thus, the operator's work is described by: $t_{n=1} = t_{3000} = t_1 . (3000)^{-b}$, and with 'b' and '$t_1$' known, the learning characteristics of this task are also known.

4.5.2 Categories 2, 3 and 4. On an entirely heuristic basis, we propose to use the power curve to provide the modifying factor for "previous experience". We first allocate an appropriate learning slope for each of our categories as shown in Table 4.12. The actual slopes were heuristically selected by observing the rate of reduction associated with each of the learning slopes and selecting the ones that seem to be 'suitable'.

Table 4.12: Learning slope "allocations" to past experience categories

Category #	Allocated Learning slope
2	70%
3	85%
4	90%

We then look at the performance reductions with each learning slope, replacing the 'number of repetitions' by the 'number of years of experience' and reading the modifying factor for "past experience" according to Table 4.13. Say for example, an operator has ten years experience in Category 4; we can then read off the table the modifying factor of '0.69'. This means that the t_1 value will be reduced to '0.69 x t_1'. A Category 2 person, with only one year's experience has an almost equivalent modifying factor of '0.70'. An example will help clarify how this information is put together.

Table 4.13: Modifying factors for past experience by category

Years of Experience.	Category #		
	2	3	4
1	0.70	0.85	0.90
2	0.57	0.77	0.85
3	0.49	0.72	0.81
4	0.44	0.69	0.78
5	0.40	0.66	0.76
6	0.37	0.63	0.74
7	0.34	0.61	0.73
8	0.32	0.60	0.72
9	0.31	0.58	0.70
10	0.28	0.56	0.69
11	0.26	0.54	0.67
12	0.24	0.52	0.66
13	0.22	0.51	0.64
14	0.21	0.50	0.63
15	0.20	0.48	0.62

With the modifying factor known, the calculations can proceed as illustrated by the example below.

Example: With a new job being scheduled, we plan to use an operator who has been with the company for five years, working, in Category 3.
   We determine the value of 'b' and 't_1' according to Section 4.2.
   Let us say, $b = 0.368$ ($\phi = 77.5\%$), and $t_1 = 4.55$ hrs.
   The appropriate modifying factor (5 yr. in Category 3), is '0.66'.
   We then find the "N" value associated with a performance time of '0.66 x t_1', i.e., '0.66 x 4.55' = 2.97 hrs. We do this by using basic equation (3.1a),

$$t_n = t_1 \cdot (n)^{-b}, \text{ putting } t_n = 2.97, \text{ and } b = 0.368, \text{ we get}$$
$$2.97 = 4.55 \, (N)^{-0.368}, \text{ whence, } N = 3.19 \text{ cycles. } (\textbf{or, N =3}).$$

Thus, the operator is still high on the learning curve, but we 'save' the time – "$t_1 + t_2 + t_3$", which equal to 4.55 + 3.87 + 3.04 = **11.5 hrs.**
   Say, our operator had 15 years experience in the same Category 3. We get a value of 0.48 from Table 4.13, and the equivalent number of repetitions he has 'completed' is N =7.3 (or, N = 7). This may not seem much, but remember, his starting time is slightly less than half of what a 'naive' operator would start at.
   Say, our operator had 15-yr. experience, but belongs to category 2. His modifying factor is 0.20 (only a fifth of t_1), and the equivalent number of repetitions completed would be 79 (or, say, 80).
   Readers are presented with a heuristic solution to the 'past experience' problem. This solution was developed during the writing of this manuscript, and therefore there is no experimental evidence to support the proposal. However, what comes out appears to be reasonable and could be tried out as an investigation. The main contribution in this section is on costing aspects. Without the heuristics, one may be obliged to assume naive operators, which will consequently drive up the costs.

5 A SUMMARY OF LEARNING MODELS WITH FORGETTING

Introduction

Most of the early research on learning curve models assumes continuous production even though interruptions occur regularly and frequently. For example, each typical daily 8-hour work shift is followed by a 16-hour break (or, interruption). Not forgetting the weekends, when a 2-day break (plus 16 hours), occurs with many workers.

Obviously, the effects of the 'daily' or 'weekend' break have not proven to be too 'damaging' to industry, else the problem would have been attended to years ago. On the other hand, the effects of forgetting should be accounted for in many other situations, especially those requiring estimates of man-hours, manpower, monitoring and control activities, rebidding repeat contracts, production planning and control, production schedules, and so on.

The degree to which an acquired skill is retained over the passage of time, is called "skill retention". Since we also consider the relearning process, we shall refer to this as the "Learning-Forgetting-Relearning (L-F-R)" phenomenon.

"L-F-R" will be discussed under the following subsections:

    5.1   Background to the "L-F-R" Process.
    5.2   How should "L-F-R" characteristics be measured?
    5.3   Factors that influence the "L-F-R" phenomenon.
    5.4   Modeling the "L-F-R" Process.

5.1 Background to the "L-F-R" Process

As expected, psychologists have led the way on research into the Learning-Forgetting-Relearning (L-F-R) phenomenon. Their research investigated the nature of memory; forgetting of 'sensible' and nonsense material; the effect of interruptions, often measured in minutes, hours and sometimes, days. However, their work appears to have little relevancy to industrial applications, though they provide a theoretical basis for later research by engineers and ergonomists. Some findings, relevant to the L-F-R area, include the following:

- Increasing previous experience reduces the effect of forgetting.
- Increasing interruption length increases the effect of forgetting.
- Task difficulty does not influence the forgetting rate.
- Slow and fast learners do not differ in their forgetting rates.
- Retraining time to achieve the original performance level is rapid.

A distinction is also made between 'procedural' and 'psychomotor' (or, 'continuous control'). Procedural tasks consist of a series of discrete motor tasks as found in most industrial activities. Psychomotor tasks, or continuous control tasks, involve repetitive movements without a clear beginning or end – such as riding a bicycle. Accordingly, the latter exhibits less forgetting than the former does.

Psychomotor skills are retained longer than procedural skills. There is major support for this in the literature (Fleishman and Parker, 1962; Goldberg and O'Rourke, 1989; Shields et al., 1979; and Prophet, 1976). Adams (1987) and Schendel et al. (1978), indicate four reasons for the difference in retention levels:

- Psychomotor skills are better organized,
- Procedural tasks include verbal components which are forgotten faster,
- The evaluation of tasks is different – it is easier to evaluate procedural tasks than psychomotor tasks. For example, you don't forget to ride a bicycle, but it is difficult to ascertain if you are riding well.
- Usually, the time required to teach psychomotor tasks is longer and as a result they are better retained.

Another distinction can be made between 'automatic' tasks and 'controlled' tasks. Automatic (or, continuous) skills are related to quick and unconscious activities (Logan, 1988; and Gopher and Donchin, 1986). Fisk, Ackerman and Schneider (1987) indicate the importance of consistency between the task components. Such consistency leads to automatism and better retention. Fisk and Hedge (1992) found there is a large decline with the performance of controlled tasks after a non-practice of one year, whereas Eggemeire and Fisk (1992) found that with automatic tasks, there is good retention even after a year. Cook, Durso and Schvaneveldt (1994) found good retention of automatic tasks after nine years.

However, we shall not consider this aspect any further in our "L-F-R" studies.

Industrial research on the forgetting phenomenon is relatively new – perhaps no more than a decade, though the 'damaging' effects of forgetting were recorded much earlier (see Keachie and Fontana, 1966; Steedman, 1970; Fleishman and Parker, 1962; Anderlhore, 1969; McKenna et al., 1985; Womer, 1979; Sule, 1978; Carlson and Rowe, 1976). Findings from these writings indicate great losses in production performances, as a function of the interruption length, as well as the experience gained.

Striking outcomes are reported, such as a 60% decay after six months and a 75% productivity loss following a 1-year break. However, these early articles lack the scientific methods needed to support such claims because of the many undefined extraneous factors (found in industry) that operate in tandem.

Does this mean an almost complete loss of a particular skill were it not used for several years? This doesn't adequately fit our real life experiences. For example, until his early twenties, this author was considered to be a moderately good swimmer. Over the years, the times between swims have extended for up to 5 years. But enter a swimming pool today, and I will swim with a fine stroke, though my lung capacity is severely diminished. There are many other examples one could find where this applies – motoring, tying shoe laces (I've not done this for nearly 40 years!), assembling shirt buttons, and so on.

How can this be explained? Is this due to 'automatism' as discussed earlier? Theories by psychologists abound – but the clue may come from the researchers themselves, since they claim that forgetting also depends on the level of learning before the break. Thus, if we are highly skilled at an activity for a sufficiently long time, then even long interruptions may not have too great an effect on performance once the activity is resumed. How would we know when experience is "sufficient" for no loss of performance skills to occur?

The importance of the L-F-R phenomenon is embedded in a vast array of real life applications, as the following scenarios describe:

- Major maintenance performed on a regular basis (weekly, monthly, etc.).
- Repairs performed on equipment failures. Average time between failures is stochastic, which can get to be quite large.
- On-line attention to automated systems – e.g., automated factories, monitoring controls in a power plant, and so on.
- Set-up for various processes. These can get to be very long (e.g., change over times for car moulds – 8 hours; set-up times for machining turbine blades – up to 2 days). The frequency of these setups may vary from several times a day (e.g., moulds used with plastic extruders) to once in several days and even longer (e.g., some car body dies are changed once a month).
- Manpower estimates for numerous situations, e.g., projects, etc.
- Mixed model assembly lines operating on a batch mode, where extensive time delays can occur between the assembly of a particular model.

- Production of very large products, such as patrol boats, tanks, aircraft, missiles, etc., where the repeat activity on a subsequent product can get to be several days, weeks, and more.
- Competitive pricing for repeat orders.
- Virtually all safety procedures – e.g., to prevent meltdown in nuclear reactors; preventing damage to expensive equipment, or, avoiding physical harm; fire drills on ocean-going vessels; earthquake drills; defense driving on icy road conditions, and so on.

5.2   How Should L-F-R Characteristics Be Measured?

There are very few research papers that have considered measures for forgetting. Published papers include works by Carlson and Rowe (1976), Bailey (1989), Elmaghraby (1990), Globerson, Levin and Shtub (1989), Shtub (1991), and Globerson, Nahumi and Ellis (1998).

Figure 5.1 shows a typical learning curve for some task when an interruption occurs at n = N. Let "L" be the break length, so that after time L, production resumes. It is expected that forgetting will occur; but as experience increases before the interruption, there is a weakening effect on forgetting.

Globerson et al. (1998) define a dimensionless number "Forgetting Parameter" as $\dfrac{t'_{N+1}}{t_N}$ (see Figure 5.1), but this does not indicate the actual

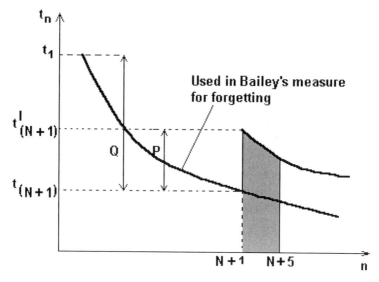

Figure 5.1: Illustrating the interruption effect

forgetting percentage that occurs. Bailey (1989) defines "forgetting" as the sum of the differences of the first four relearning observations from their respective points on the original learning curve as illustrated in Figure 5.1. While this definition may have been useful for Bailey in his innovative study, his relearning data is confounded by the summing process and important information on the four individual points (but especially the first one) is lost.

Figure 5.2 illustrates Carlson and Rowe's (1976) approach to the L-F-R process. This figure uses a "Performance – Time" relationship which contains the same information as does Figure 5.1. However, the break length can be drawn to scale in this graph. Incidentally, their proposed solution fixes the Learning and Relearning slopes at 87%, while they assume that Forgetting can be represented as a power decay function whose learning slope is 80%. Their work was published some 25 years ago and represents an innovative, though limited approach for tackling the L-F-R problem.

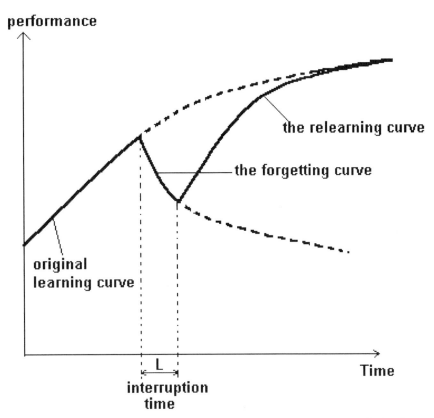

Figure 5.2: Learning-Forgetting-Relearning on a 'performance v/s time' graph

Returning to Figure 5.2, one can add an asymptotic value of the 'maximum performance' that would represent a 'fully learned (or, trained)' condition. If subjects are expected to achieve this level, then deviations from this level truly represents how much learning is lost. In a very recent article, Globerson et al. (1998) do just that. In their investigation, experienced pilots were asked to practice on their assigned task "....until their performance results reached a plateau". A similar piece of research was done by Ginzburg (1998), who required subjects to reach some predefined performance level (discussed in Chapter 7, Section 7.2).

Referring to Figure 5.3, Globerson et al. (1998) propose a new definition for "forgetting" as follows:

- Absolute Forgetting (AF) = P   (5.1)
- Relative Forgetting   (RF) = P/Q   (5.2)

Both these measures make sense and are adopted as the standard measurement for forgetting in this book.

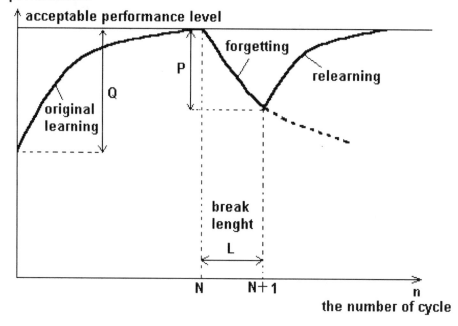

Figure 5.3: Forgetting with respect to some 'acceptable performance' level

We now consider the factors that influence the L-F-R process.

5.3   Factors that Influence the "L-F-R" Phenomenon

It is most likely that the factors influencing individual learning (Chapter 2) will also influence the L-F-R process. In this section, we discuss factors that specifically influence L-F-R for individuals. Learning and forgetting have also been written about in organizational settings and this aspect will be discussed in Chapter 8.

The following factors are not listed in order of importance, but are presented in a logical order:

5.3.1   The Break Length
5.3.2   Previous experience (the amount of learning prior to the interruption)
5.3.3   Job complexity
5.3.4   The work engaged in during the interruption period
5.3.5   The cycle time of the task
5.3.6   The Relearning curve
5.3.7   When Relearning is a single observation point.

5.3.1 The break (or, interruption) length. Forgetting as a function of the break length, L, has been known for years, beginning with the work by psychologists (see, Klatzky, 1980; Lanchman, 1979; Kyllonen and Alluisi, 1987; Hulse, 1975; Ekstrand, 1967; Heally et al., 1995; Underwood, 1954, 1968; Wickelgren, 1977, 1981; Christiaansen, 1980; Carron, 1971; Schendel et al., 1978; Hagman and Rose, 1983). Engineers and ergonomists have also done their part (see Steedman, 1970; Keachie and Fontana, 1966; Anderlhore, 1969; Cochran, 1968; Sule, 1978; Venda, 1995; Wisher et al., 1991; Carlson and Rowe, 1976; Bailey, 1989; Smunt, 1987; Baloff, 1971; Yelle, 1979; Asseo, 1988; Dar-El et al., 1995; Sabag, 1988; Globerson et al., 1987, 1989; Arzi and Shtub, 1997; Vollichman, 1993; Livne and Melamed, 1991; Sparks and Yearout, 1990; Hewitt et al., 1992).

While these researchers have considered forgetting with respect to break length, the work by Globerson et al. (1998) is singled out as providing outstanding data on this topic covering both cognitive and motor tasks. Their results on recording the Relative Forgetting (RF) from equation (5.2) is given in Table 5.1 and graphed in Figure 5.4.

Figure 5.4 is typical of 'RF v/s L' relationships for a specific amount of previous experience, and is consistent with the results obtained by recent researchers (see Asseo, 1988; Globerson et al., 1989; Shtub et al., 1993; Arzi and Shtub, 1997; Vollichman, 1993).

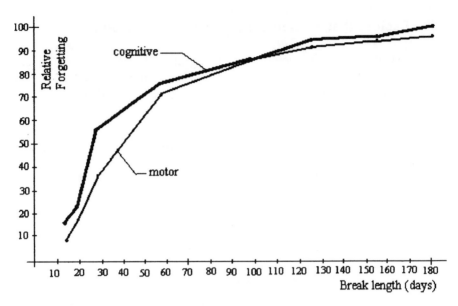

Figure 5.4: Globerson et al. (1998) data graphed

Table 5.1: Relative Forgetting (RF) for Globerson et al. data (%)

BREAK LENGTHS (days)

Task	14	21	28	58	98	127	154	180
M*	7	17	36	69	84	90	91	93
C	14	21	56	74	84	93	93	96

*M - Motor task   C - Cognitive task

Note: It's a moot point how this task was defined as 'motor'. Even the authors admit to this. A pure motor task would exhibit a much wider gap than the cognitive graph. Indeed, the evidence shows that pure motor exhibits no forgetting whatsoever (see Bailey, 1989; Cooke et al., 1994; Dar-El et al., 1995), in which case, the graph in Figure 5.4 would be a horizontal line with RF = 0.

Data from Vollichman (1993) is based on similar experiments and RF values are included in Table 5.2. Vollichman's experiments dealt with a simulated chemical plant in which large containers (storage tanks) were required to be filled with an unstable inflammable fluid. Overfilling the containers could result in an explosion and certain courses of action were defined in order to avoid an explosion from occurring. The subjects were asked to operate the simulation model in one of three modes:

- H – Maximum attention to quality – but not to neglect speed
- M – Medium attention to quality, with more attention to speed
- L – Maximum attention to speed, but to watch out for potential explosions.

These three levels are somewhat equivalent to three levels of 'job complexity'. Measured learning slopes were: 83.5% (H), 77.4% (M), 72.5% (L).

Table 5.2: Relative Forgetting (RF) for Vollichman's data (%)

BREAK   LENGTH (days)

Performance Quality	1 week	2 weeks	1 month	2 months
H	33	40	58	84
M	21	29	40	62
L	19	20	31	45

The general trends for RF in Tables 5.1 and 5.2 are consistent in that RF increases with the break length. Comparing the two tables, the differences in RF values are a lot larger for smaller breaks than for the longer breaks (> 2 weeks). Clearly, job complexity plays a major role in determining RF values (note, the psychologists claimed there was no linkage between the task difficulty and forgetting). The data for these five 'conditions' provide a good starting point for obtaining initial values for RF for predictive purposes in various situations.

In Figure 5.4, operators were asked to practice until their performance times plateaued before the interruption was applied. Yet in the studies with Vollichman (Table 5.2), the subjects were clearly in the learning phase. How RF values would behave with respect to previous experience will be the next factor to be discussed.

5.3.2 Previous experience. As with the break length, previous experience is an important factor in the level of forgetting. Few researchers were involved in this aspect of 'forgetting' research, and even fewer have found their way in scientific publications (see Dar-El et al., 1995; Vollichman, 1993; and Dar-El and Zohar, 1998).

A typical learning curve with several breaks is illustrated in Figure 5.5. The influence of previous experience is seen to diminish as experience is gained. This also depends on the job complexity; if this were high, the effects would be more severe than if the tasks were less complex. For largely motor tasks, one can safely assume that the relearning 'pips' on the learning curve would be sufficiently small to be ignored altogether.

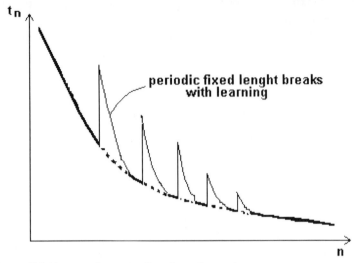

Figure 5.5: Interruptions as a function of experience

The expected form of the relationship between RF and previous experience should resemble the graphs in Figure 5.6.

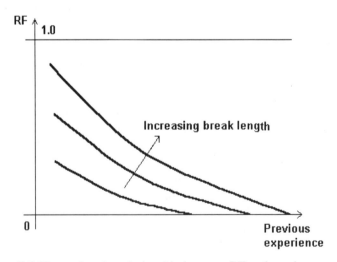

Figure 5.6: Illustrating the relationship between RF and previous experience

Table 5.3 includes the full set of RF data obtained from Vollichman's experiments, since these include breaks of: 1 week, 2 weeks, one month and

2 months, respectively. The 1 and 2-week break study was repeated 4 times; the 1-month break was done three times and the 2-month break was repeated twice. Each L-F-R cycle involved 16 repetitions of the task. Table 5.3 clearly shows RF values reducing with experience. The categories of "H", "M" and "L" in Table 5.3 refer to the quality level required: "H" means High; "M" means Medium, while "L" means Low quality requirements.

Table 5.3: RF values for Vollichman's experiments

	BREAK LENGTHS FOR "H", "M" and "L"											
L-F-R set #	H				M				L			
	1w	2w	1m	2m	1w	2w	1m	2m	1w	2w	1m	2m
1	33	40	58	84	21	29	40	62	19	19	31	45
2	23	35	56	85	19	26	40	58	13	18	25	41
3	23	29	52	-	17	22	38	-	14	16	22	-
4	23	30	-	-	12	20	-	-	12	14	-	-

5.3.3 Job complexity. Job complexity was already shown to have a major influence on individual learning (with no forgetting). It also appears to have an influence on the 'forgetting' characteristics. We have already pointed out the supporting evidence shows that well learned 'highly motor' tasks (ϕ around 90%) exhibit negligible forgetting since there is very little to forget (see Fleishman and Parker, 1962; Bailey, 1989; Cooke et al., 1994; and Dar-El et al., 1995). However, this characteristic is not observed in Globerson et al. (1998), whose RF data is shown in Table 5.1 (and Figure 5.4). Indeed, their two graphs tend to merge as the break length increases. Analyzing the description of the work done by pilots in their experiments, it would appear that what they call "motor" tasks, still have plenty of cognitive elements embedded in these activities, and consequently, it is not surprising that their RF data exhibits 'real' forgetting for their motor tasks.

In any case, one does not usually find pure cognitive or pure motor tasks in industry. What we do find are tasks that range from 'highly cognitive' (ϕ about 73%) to "highly motor" (ϕ about 88%), with each extreme containing a bit of the other (see also Bailey, 1989; McKenna and Glendon, 1985; and Hewitt et al., 1992).

The tendency in industry is to automate mechanical tasks as much as possible and to utilize people for control and decision-making purposes. A lot more research is required to link job complexity (especially 'highly cognitive' tasks) with break length and previous experience in order to obtain a better understanding of the 'forgetting' process.

5.3.4 The work engaged in during the break period. Not much work has been done in this particular area. Ekstrand (1967) and Heally et al. (1995) observed that forgetting was less if subjects rested (including sleep) during the break period, than if they were busy with other activities. But Hovland (1951) suggests that if the work performed during the interruption was of a similar nature, then the forgetting rate could be reduced. Arzi and Shtub (1997) found strong evidence supporting the claim that performance improvements occur *after* the break (i.e., compared with *before* the break). Their experiments required the solution to complex pattern recognition problems. This apparently requires a mental selection of one of several algorithms for solving the problem. They suggest that some subjects were sufficiently motivated to sharpen their algorithmic skills during the break in order to improve on their respective performances. This happened with 30% of their subjects and this point cannot be ignored. Apparently the 'others' did not go through this process and, as expected, exhibited clear forgetting characteristics. The possible implications to educational programs could be highly relevant and obviously call for more research to explore this very interesting phenomenon.

In terms of industrial applications, one has no control of what the worker does once he, or she, leaves the plant, and as a consequence, work done during the break period cannot be considered as an independent factor.

5.3.5 The cycle time of the task. Dar-El et al. (1995)[*] were the only researchers to have considered 'forgetting' as a factor when determining the learning characteristics of long cycle time tasks. As early as 1962, Kilbridge reported extensive learning studies done in the electronic industry, where MTM cycle times ranged from a few seconds to about ten minutes. Kilbridge observed that the maximum performance rate occurred for cycle times between 1/3 min. to 1½ min. Thereafter, the performance rate decreased, with 'forgetting' cited as the most likely cause for causing the decrease. Thus, the larger the cycle time (beyond 1½ min.), the poorer the maximum achieved performance rate.

Sabag (1988) tested this assumption in a cleverly designed experiment which used a task comprised of three distinct steps, 'A', 'B' and 'C'. One set of subjects was asked to use Method 1, which assembled the whole task (i.e., A + B + C) twenty times. The second set of subjects used Method 2, in which the first sub-task, 'A', was done twenty times, followed by the second sub-task, 'B', done twenty times, followed by sub-task 'C', done twenty times (i.e., 'A' x 20, followed by 'B' x 20, followed by 'C' x 20). The total

[*] Use of copyright material by *IIE Transactions* is gratefully acknowledged.

MTM for "A + B + C" was just under three minutes. The following results were the average times, using seven subjects for each method:

Method 1 ⇒ 258.0 minutes,
Method 2 ⇒ 213.7 minutes.

Differences were statistically significant at the 0.025 level and were most likely explained by 'forgetting' occurring between each cycle when using Method 1. Thus, on completing 'A', the operator proceeds to complete 'B' and 'C'. The time for completing 'B + C' was actually an interruption period for sub-task 'A'. The same argument would apply for each sub-task 'B' and 'C'.

Dar-El et al. (1995) proposed to use this approach for tasks whose MTM times ran to several hours, or even days. The model they proposed assumed that a long cycle time task is comprised of a series of unique sub-tasks such as that illustrated in Figure 5.7. The average MTM times for these sub-tasks are about 1½ min. and the assumption is made of sub-task independence with no transfer of skills between them.

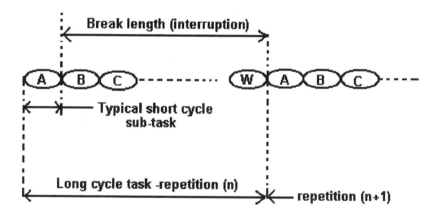

Figure 5.7: A model for long cycle tasks

This model was derived from observing long cycle time work being performed in several R&D companies. Operators would move from one sub-task to the next, defining (to themselves) the next objective to be accomplished in a specific sequence. Most often, there did not appear to be any commonality in activities performed between adjacent tasks, although, considering the whole cycle (of the long task), some repetition of simple sub-tasks did occur.

Once a sub-task is completed, it is assumed to appear again only in its correct sequence on the next cycle. Thus, the repetition of the task is interrupted by a time approximated by the cycle time. This interruption causes 'forgetting' to occur, by an extent that is clearly a function of the interruption length, or, the cycle time itself.

Thus, 'forgetting' with long cycle time tasks, is treated as a forgetting problem. The actual development of predictive equations is treated in Chapter 6, Section 6.3 which covers 'applications to assembly systems'.

There is no research material available that considers the task 'cycle time' as a factor in the L-F-R process. One can surmise that if job complexity of a long cycle task is increased, then this could further worsen the forgetting characteristics of an interruption, but this aspect still requires investigation.

5.3.6   The relearning curve. From observation made on the results of many experiments (see Asseo, 1988; Carlson and Rowe, 1976; Livne and Melamed, 1991; Bailey, 1989; Shtub et al., 1993; and Dar-El et al., 1995), one can conclude that the relearning curve rapidly approaches and merges with the original power graph of the learning curve. Furthermore, the power model makes a good fit with relearning data. However, there is no exact relationship between the learning slopes of the original power curve and the relearning curve. Indeed, the slope of the relearning curve may even be higher (i.e., slower learning) than the original learning curve, but because relearning can be considered as if the number of cycles begins from n = 1, 2, 3,..., there is a very rapid improvement in the early cycles, which most often smoothly merges into the original learning curve, as illustrated in Figure 5.8. In the work by Bailey (1989), Dar-El et al. (1995) and Globerson et al. (1998), the relearning learning rate is smaller (i.e., faster learning) than the original learning curve.

On the other hand, when interruptions are especially long and previous experience limited, there could be a relatively large relearning cost involved. Globerson et al. (1998) calculates this cost as the sum of the shaded area in Figure 5.9. They call this the recovery Cost (RC) which is expressed as follows:

$$RC = \int_{N+1}^{N+T} Ti - \int_{N+1}^{N+T} Ti \qquad (5.3)$$

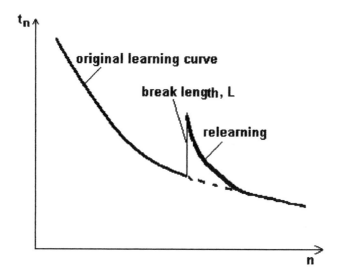

Figure 5.8: Illustrating the characteristics of the relearning curve

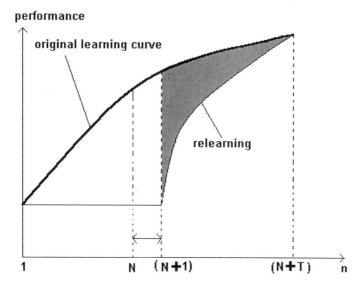

Figure 5.9: Determining the cost of forgetting

5.3.7   When relearning is a single observation point. A number of real life situations occur when relearning is a single performance of the task. For example, virtually all maintenance operations fall into this category. An operator is asked to do repairs either during a planned maintenance period (e.g., once in three months) or else, following an equipment breakdown.

The repair, or maintenance, is done just once until the next maintenance period, or, the next breakdown. This condition is illustrated in Figure 5.10, which shows each 'relearning' as a single observation. Without a doubt, learning and overall retention occurs.

Another example is the response needed to counteract a safety emergency. For example, if the pressure in a vessel becomes too high, a counteraction is needed by the operator in order to avoid an explosion. The times between "failures" are usually stochastic and operators must be trained to deal with such emergencies.

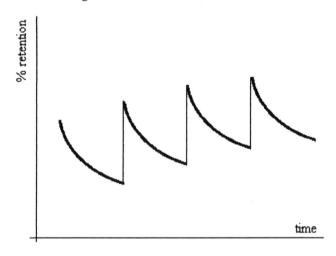

The case of 'single' relearning
(e.g., maintenance)

Figure 5.10: "L-F-R" for single observations (e.g., maintenance)

Section 5.3.5 dealing with "long cycle times", is yet another example of a single observation point followed by an interruption period, except in this case, the 'single point', occurs on a regular basis with each new product cycle time (this case is discussed further in Section 6.3).

Apart from the application to long cycle times, no research exists for dealing with 'relearning as a single observation point'. In the earlier two illustrations (with maintenance, and counteracting emergencies), the times between calls for action can get to be very large (months, or years) and operators are likely to forget everything they had learned from the previous maintenance (or, emergency) action. In order to maintain some 'memory' of these activities, operators are put through some simulated training programs, which help retain knowledge of the tasks. Combat fighter pilots and navigators do this retraining on a routine basis in order to maintain a high degree of their respective skills. Fire drills on passenger vessels or, in high-rise hotels, are useful; as are off-site practice drills for set-up operations and maintenance.

Recording the procedures taken and reviewing these on a regular team basis, also helps in refreshing the memory in such situations. Dar-El and Zohar (1998) were involved in determining optimal training schedules for tackling emergency actions. The results of their work will be discussed in Chapter 7.

5.4 Modeling the "L-F-R" Phenomenon

It is clear that the "L-F-R" phenomenon falls into two distinct categories:
  (One)   when 'Learning' and 'Relearning' include multiple repetitions;
  (Two)   when 'Learning' and 'Relearning' are single repetitions (as in maintenance).
An example of (b) is discussed in Section 6.3, but other than this, *no* other research has been reported, and consequently, category (b) will not be discussed further – we await future research! This section will only deal with category (a), where 'Learning' and 'Relearning' include multiple repetitions.

5.4.1 Modeling "L-F-R" for multiple repetitions. Very little research has been done in this area. Carlson and Rowe (1976) proposed a general model based on several restrictive assumptions, while Globerson et al. (1989) and Shtub et al. (1993) both use the same data base (the work of Asseo, 1988) to develop models that are specific to the area of Asseo's work. The latter's experimental work dealt with data entry into a computer – 16 cycles followed by varying break lengths, followed by another 16 cycles of data entry. Their respective models are narrowly defined, predicting data entry times when the interruption occurs after 16 cycles, followed by varying break lengths, and are of little use to industry.

A major 'push' in understanding the L-F-R process is undoubtedly the innovative work by Bailey (1989) and Globerson et al. (1998); though others too have made their contribution (see Globerson et al., 1987, 1989; Vollichman, 1993; and Arzi and Shtub, 1997).

Summarizing our knowledge thus far, we can relate to the three components of the L-F-R process:
  (One)   The initial learning curve – no problem here. Chapter 3 deals with several learning models, and we can estimate t_1 the first

cycle time, as well as the performance time when the interruption occurs.

(Two)   The forgetting period – clearly a function of the previous experience as well the interruption period. We now have data for some five examples from which RF (Relative Forgetting) factors are known. We can extrapolate between these to help select a RF appropriate to our current needs.

(Three)   The relearning period – we conclude that the power model nicely fits the relearning data and that relearning is very rapid, allowing the performance to meet up and merge with the original learning curve after a few cycles.

We are now in a position to propose a 'L-F-R' model.

5.4.2 Proposal for a "L-F-R" model. We propose a model quite similar to that used by Carlson and Rowe (1976), but without the inclusion of their restrictive assumptions. Their model description was shown in Figure 5.2 and is reproduced as Figure 5.11, but with the abscissa represented by "the number of cycles". The graph in Figure 5.11 is comprised of the three L-F-R components; 'A B' representing the original learning curve, 'B C' the decay (or, forgetting) curve, and 'C D', the relearning curve.

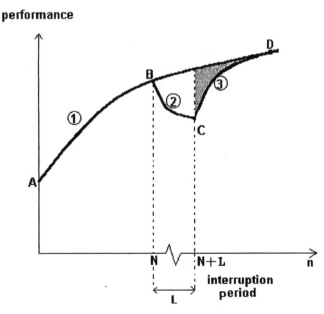

Figure 5.11: A Learning-Forgetting-Relearning model

Referring to Figure 5.11,

Curve "A-B"   should present no problem. We can now find points 'A' and 'B'.

Curve "B-C"   needs data on the RF (Relative Forgetting) value. Until more information becomes available, we use Tables 5.1, 5.2 and 5.3 to extrapolate an appropriate RF value for our specific case. We now find point 'C'.

Curve "C-D"   can be estimated by assuming that the relearning slope is identical to the slope of the original learning curve. The justification for this assumption is that 'recovery' is rapid and most studies show the relearning slope to be slightly smaller (i.e., faster learning) than the learning slope of the original curve. Thus, the damage is likely to be relatively small and on the conservative side. This gives us point D.

This model enables points A, B, C, and D to be evaluated, and the Recovery Cost (RC) can be calculated using equation (5.3).

5.4.3 An example. Consider that Curve "A B" represents the completion of a contract to supply 100 high precision items to a customer. Some 3 months later, the customer asks the plant for a new quote for another 100 items – indicating that a lower quotation is expected since 'learning' had occurred.

Your task is to provide your manager with all the information needed to enable a decision to be made.

Looking up plant records, you note that the work was performed by a 2-man team and that $t_1 = 6.5$ hours, and $\phi = 82\%$.

Man-hours invested for producing the first order of 100 items:
   From equation (3.7), we have

$$T_{100} = \left[t_1 \cdot 100^{(1-b)} \right] 1/(1-b)$$

$t_1 = 6.5$ hrs and b = 0.286 (from $\phi = 82\%$, and $\phi = 100\,(2)^{-b}$). Substituting into the above, we get

$$T_{100} = 1/(1-0.286)\,\{6.5\,.100^{-0.286}\} = 243.9 \text{ hrs.}$$

With two workers, man-hours invested = 487.8 hrs.

Theoretical man-hours invested for the 2nd order with no forgetting:
   Man-hours needed to produce 200 items are:

$$T_{200} = 1/(1-0.286)\left\{6.5\cdot 200^{-0.286}\right\}= 400.1 \text{ hrs.}$$

With two workers, man-hours invested = 800.2 hrs.
Hence, Man-hours needed for the 2nd order = 800.2 – 478.8 = 312.4 hrs.

To find Point "C" in Figure 5.11:
    Refer to Tables 5.1, 5.2 and 5.3. For previous experience of 100 items, we must turn to Table 5.3. High precision means that we should look under the "H" column. The fourth row of Table 5.3 represents a past experience of 64 items and no data is available for even the 2-month break. For convenience, we reproduce a part of Table 5.3 below:

L-F-R cycle	1 week	2 weeks	1 month	2 mths.	3 mths.
1	33	40	58	90	
2	25	35	56	90	
3	23	29	52	NA	
4	23	30	NA	NA	
5					
6	18	24	45	65	70

    Our first estimates for row 6   (representing 96 repetitions) gives the underlined values. Of particular interest is the last entry of 70% (which is extrapolated).
    Thus, RF = 70%
    Point "B" (Figure 5.11) is, $t_{100} = 6.50 \cdot 100^{-0.286} = 1.74$ hrs.
    From Figure 5.1, Q = (6.50 – 1.74) = 4.76 hrs.,
    and from equation (5.2), P = 4.76 . 0.70 = 3.33 hrs.
Therefore point "C" = 3.33 + 1.74 = 5.07 hrs.

To find Point "D" in Figure 5.11:
    Let "x" be the number of cycles where the two graphs meet at Point "D". For the original curve, $t_X = 6.50 \cdot (x)^{-0.286}$, and for the relearning curve, $t_X = 5.07 \cdot (x-N)^{-0.286}$, i.e., $6.50 \cdot (x)^{-0.286} = 5.07 \cdot (x-N)^{-0.286}$ taking logs –

$$\log (6.5/5.07) – 0.286 \log (x) = - 0.286 \log ((x-100)$$
$$\text{i.e.,} - 0.3773 + \log (x) = \log (x-100).$$

Through a process of iteration,   x = 172 hrs.

To find the excess manpower due to forgetting:
    (i.e., the shaded area in Figure 5.11)

$$T_D = T_{172} = (1/(1-b)) \cdot [t_1 \cdot (172)^{(1-b)}]$$
$$= 1/0.714 [6.5 \cdot (172)^{0.714}]$$
$$= 359.3 \text{ hrs.}$$

Therefore the area under the original curve between B and D is,

$$(359.3 - 243.9) = 115.4 \text{ hrs.},$$

and the area under the relearning curve between B and D is,

$$T_{72} = (1/(1-b)) \cdot [t_1 \cdot (72)^{(1-b)}]$$
$$= 1/0.714 [5.07 \cdot (72)^{0.714}]$$
$$= 150.5 \text{ hrs., and using 2 workers give, 301 hrs.}$$

Therefore the shaded area in Figure 5.10, or, the Recovery Cost, RC is

$$RC = 150.5 - 115.4 = \underline{35.1 \text{ hrs.}}$$

Using 2 workers, this gives 70.2 hours. Thus, the total hours to make the second order of 100 items would require 312.4 + 70.2 = 382.6 hrs (compare this with the 487.8 hrs needed for the first order).

5.4.4 A summary. The weakest part of the model is in finding the RF factor. We will be on firmer ground when more data of the type shown in Figures 5.1 to 5.3 become available. Perhaps the RF factor can be replaced by a simpler power representation for the forgetting curve expressed as a function of the break length and previous experience, but this will need to be investigated.

The "L-F-R" process for multiple repetitions has been covered to the best extent possible given the current research. Obviously, there are numerous areas that need to be further researched in order to fill the many blank spaces in our current knowledge of the topic. One purpose of this chapter was to identify the particular needs for future research in this area.

6 APPLICATIONS (WITH & WITHOUT FORGETTING)

Chapters 2 to 5 have mainly dealt with the theoretical side of learning curve theory. This chapter is devoted to "Applications", but as far as possible, we shall concentrate on research-type literature that deal with applications to specific areas.

Applications of Learning Curves (LC) will be discussed as follows:

6.1   Learning Curves in Time and Motion Study
6.2   Learning Curves in Assembly Line Design
6.3   Learning Curves with Long Cycle Time Tasks
6.4   Learning Curves in Batch, or, Lot Size Production
6.5   Learning Curves in a GT (Group Technology) Environment
6.6   Learning Curves in a JIT Environment
6.7   Speed – Accuracy in a Learning Environment.

There are a vast number of other potential application areas, but research literature is lacking.

6.1 LC in Time & Motion Study

Learning Curves are very basic to the Time & Motion Study (or, Work Study), since much of the early work on this subject dealt with task performance times and their improvement (see any basic book on Time & Motion Study, e.g., Barnes, 1980; Mundel and Danner, 1995; Konz, 1995; Neibel, 1976, etc.).

I also include my knowledge on this topic, observed during my many consulting forays in industry. The best example was the use of LC in a metals processing plant in Ashkelon (Israel), where an 80% learning curve was used for establishing dynamic time standards for individual wage incentive purposes (Dar-El, 1990) as illustrated in Figure 6.1. The stepped graph is an upper bound (envelope) of the actual learning curve.

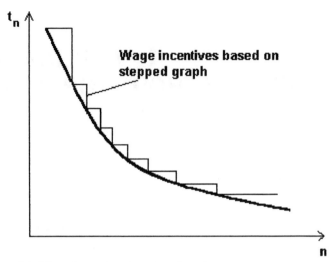

Figure 6.1: Wage incentives based on learning

The width of the steps was made slightly flexible to accommodate job complexity, but Time Standards were continually pushed down and the production department was run very efficiently (see also Finn, 1984). This contrasts with *most* plants in developed countries around the world, where LC are not taken into account. Operators put great pressure on production management to establish Time Standards as early as possible in order to benefit by the additional incentive bonus earned from expected improvements due to learning.

A large R&D plant producing electronic-based products having long cycle times, employs a cruder system for production planning and incentives. It uses the average product completion time for the just completed month as the Time Standard for the subsequent month. Since production quantities are never very large, operators have a 'ball'—deciding how much incentive bonus they should earn for the month.

In a paper written many years ago, Barany (1982) proposed that 'learning' be used for modifying Time Standards. The experiment involved students (50:50 male & female) working on a simple pattern recognition task, 2 hours a day for 15 days. Part of his data was presented as a graph which is reproduced as Figure 6.2, which clearly shows that

  a)   Learning *does* occur
  b)   Daily breaks (8 hrs) do cause forgetting
  c)   Relearning is rapid
  d)   The one male dominated the performance of the female.

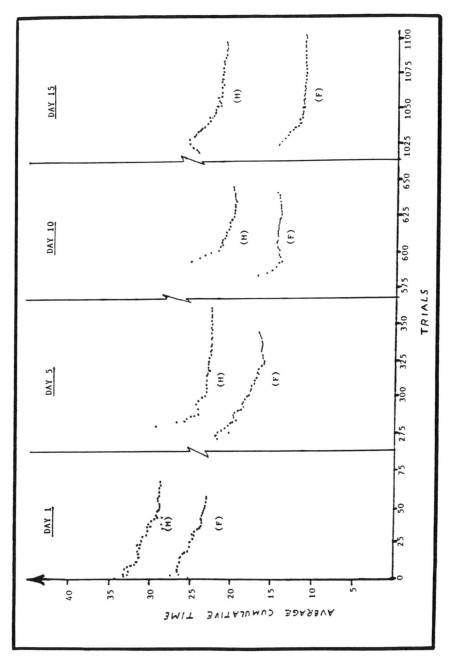

Figure 6.2: Learning curves for one male and one female subject performing the experimental task for fifteen days (from Barany, 1982)

In another paper, Globerson and Seidmann (1988) investigate the dynamic relationship between motivational techniques, such as goal-setting, and incentive payment systems versus performance improvements, as depicted by learning curves (see also Globerson and Riggs, 1988). They use the term "dynamic relationship" since it expresses the Time Standard as a rate of expected improvements. The laboratory project involved parts being fed to the operator under paced and unpaced conditions. In effect, the authors were comparing differences between paced systems (where the conveyor speed was reduced according to their measured learning slope, after every two parts were delivered) and unpaced conditions - where operators set their own natural pace for the assembly together with a good incentive plan. They also considered whether operator training in methods engineering using unpaced conditions could improve the learning curve. Table 6.1 summarizes these results.

Table 6.1: Results of Globerson and Seidmann's (1988) Experiments[*]

Group #	Assigned Slope	Actual Slope	SD of Slopes	R^2	Comment
1	-	94.9	2.0	0.84	Self-paced, no incentives.
2	92%	97.8	2.4	0.66	Imposed learning 92% slope
3	-	92.6	3.1	0.76	Unpaced –amplified incentives
4	-	92.9	2.7	0.81	Unpaced, amplified incentives method improvement training

The major conclusion is that unpaced conditions with incentives are superior to paced conditions. Furthermore, if the paced conditions were perceived as being too tough, then subjects simply stopped making the effort, no matter what the loss in income. This is true for incentive systems with 'tight' standards, even when the work rate is unpaced. If operators cannot earn much even after considerable effort, they simply stop cooperating!

That unpaced conditions with incentives are superior to paced conditions, confirm our experience with assembly lines, where it is known that unpaced conditions with incentives are superior to paced assembly lines, even though production managers find paced conditions more 'appealing', since, at least, it guarantees the production rate.

---

*Use of copyright material by *IIE Transactions* is gratefully acknowledged.

Globerson and Seidmann's example is most unfortunate, since even under unpaced conditions, the learning curve falls smack into the "High Motor" category, where there just isn't much to "learn"!

6.2 LC in Assembly Line Design

One would generally expect learning to have a major impact on production design and flow line scheduling in particular (Wilhem and Sastri, 1979; and Chen, 1983). But instead, assembly line design turns out to be one of the main application areas for learning curve theory. Research papers deal with several aspects of assembly design and will be discussed under the following subtitles:

6.2.1 The Optimal Number of Stations under Learning
6.2.2 Minimizing the Makespan for Assembly Lines under Learning
6.2.3 LC with Mixed-Model Assembly Lines
6.2.4 LC in Assembly Lines for New Products
6.2.5 Optimizing "Cycle Time : Station Number" for Assembly under
        Learning

6.2.1 The optimal number of stations under learning. Assembly lines are used for progressively building a product as it passes the stations in a sequential manner until completion at the last station. Normally, assembly lines were built to last for months, producing the same products (some with minor variations) in large quantities. Today, however, assembly lines are often used for batch production, in which an order may be completed in a few days.   Line designers know the headache of start-up conditions (Cochran, 1969; and Pegels, 1969). This is typical of complex products, partly because the conditions for the first cycles are still being learnt and coordinated among the stations comprising the assembly line. More than that, small orders for new products using assembly lines (as in high-tech plants) are invariably working under pressure in order to complete the work by the delivery date. The main culprit is "learning". Line designers do have an idea that learning takes place, but apparently, the great majority do not know how to evaluate, or plan for this condition.

There are at least two approaches for tackling this problem. One is to begin the assembly with *more* stations than would normally be required; then, as learning takes place, to begin reducing the number of stations, one at a time, until we get to the level one would want to use if there was no learning. However, this is a difficult solution to implement, since we first need to determine the number of stations with which to begin. Then, as each station is removed, the tasks in that station need to be allocated among the remaining stations, and the assembly line needs to be rebalanced. Many of these stations will now have new tasks added to their workload, which the will be working on for the first time, i.e., high up on the learning curve.

The second approach is more elegant. Taking learning into account, it uses a cost minimization approach to optimize the number of stations required that can meet specific due dates. In a paper by Cohen and Dar-El (1998), the assumption is made that all stations will be balanced and that all station learning slopes have the same value (this assumption is quite reasonable since methodologies exist that can do just that).

Consider an assembly (or, production) line having "L" stations as shown in Figure 6.3.

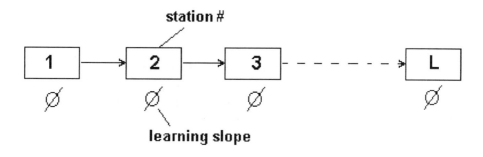

Figure 6.3: A L-station assembly line with learning

If all stations are balanced and have the same learning slope ø, then the following factors can be evaluated:

a)   The Time Standard (STD) – value of the whole assembly task
     (e.g., use MTM, MOST, etc.)

b)   The allocation to each station would be $\dfrac{STD}{L}$ and this is the standard

     cycle time, C (i.e., $C = \dfrac{STD}{L}$).

c) The learning slope 'b' and time for the first cycle for each station
""t_1 using methods from Chapter 4. The makespan (M_{max}) is then
determined by two components:
  (a) the time taken for the first product to go from Station 1 to
      Station (L-1) (see Figure 6.3);
  (b) plus the time taken for the order size, B, to be processed at the
      last station.
The first component is simply "$(L-1) \cdot t_1$" since, from equation (4.4),

$$t_1 = \frac{STD}{L} \cdot \left(\frac{t_1}{C}\right) \tag{6.1}$$

$$(L-1) \cdot t_1 = \frac{(L-1)}{L} \cdot (STD) \cdot \left(\frac{t_1}{C}\right) \tag{6.2}$$

The second component is simply T_B, the cumulative assembly time for B
items, where

$$T_B = \frac{t_1 \cdot (B)^{1-b}}{1-b} = \left[\left(\frac{STD}{L}\right)\left(\frac{t_1}{C}\right)\frac{1}{(1-b)}\right] \cdot (B)^{1-b} \tag{6.3}$$

Thus, the makespan, M_{max}, is the sum of equations (6.2) and (6.3), i.e.,

$$M_{max} = \frac{(L-1)}{L} \cdot \left(\frac{t_1}{C}\right) \cdot STD + \left\{\left(\frac{STD}{L}\right)\left(\frac{t_1}{C}\right)\right\}\left\{\frac{(B)^{1-b}}{1-b}\right\} \tag{6.4}$$

After several simple mathematical steps, we arrive at a value for L, as

$$\frac{\left\{\frac{M^{(1-b)}}{(1-b)}\right\}-1}{\left\{\frac{C_{max}}{STD \cdot \left(\frac{t_1}{C}\right)}\right\}-1} L = \tag{6.5}$$

We can now optimize L for a given due date (DD).

Assign initial values to index i = 1,2,...L. According to equation (6.5),

    a)   Set the number of stations at $(L-i)$.

    b)   Calculate STD and C, the work allocated to each station.

    c)   Calculate "b", "t_1" and find $\left(\dfrac{t_1}{C}\right)$ from Figure 4.2.

    d)   Calculate the makespan from equation (6.4).

    e)   If the makespan from (iv) is less than the due date, DD, put $i = i+1$ and go to step (i)

Else, put $i = i-1$, and the minimal number of stations for a given DD is $L-i$., i.e.,

$$(L)_{optimal} = (L-i) \tag{6.6}$$

Cohen and Dar-El (1998) also consider finding $(L)_{optimal}$ under profit maximization approach, but this adds nothing to the learning implications and interested readers can review the original paper for further details.

6.2.2 Minimizing the makespan for assembly (production) lines under learning.

The general assumption made, or implied in assembly line balancing methodology (e.g., see Talbot et al., 1986; Baybars, 1986; and Dar-El, 1991) is that station learning is nonexistent. Current research (see Goralnick, 1998) indicates that it is insufficient to merely obtain equal balance for all stations, but that the station learning slopes should be equal as far as possible. To achieve this is no great problem and several efficient heuristic methods are available to give such balances. However, a situation may arise (for several reasons) when it is *not* possible to balance the learning slope between stations. Under these circumstances, it is *not* a good idea to balance the workload between stations, as we will see below.

Consider two adjacent stations (A feeding into B), where $\phi_A < \phi_B$ (i.e., Station A learns faster than Station B). It was shown (Cohen, 1992) that in order to find the optimal makespan, the work allocation to Stations A and B must be such that an intersecting point occurs, with the respective learning curves, during the production period, as illustrated in Figure 6.4. It is clear that the allocation to Station A is greater than to Station B, and that Station B begins its first operation when Station A begins its second cycle.

It can be deduced from this simple illustration that the upper envelope in Figure 6.4 represents the actual production rate for products leaving a 2-station line. At the beginning, Station B has a shorter cycle time than Station A, and must therefore wait at each cycle for the product to arrive from Station A. After the intersection point, the first station has a shorter cycle time than Station B and its finished parts either enter a buffer before Station B, or, if the buffer capacity is zero, Station A will be blocked until Station B completes its work on the product.

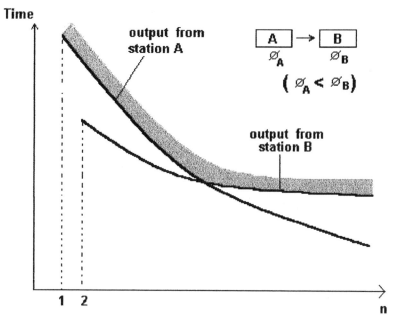

Figure 6.4: A 2-station assembly line with $\phi_A < \phi_B$

We can look upon the upper envelope of Figure 6.4 as the learning curve of a combined station (A and B), to which, may be added a third station, C, whose $\phi_C > \phi_B$, so that another intersection between the two curves occurs as in the earlier case.

This analysis was based on $\phi_1 < \phi_2 < \phi_3 < \ldots < \phi_L$ but would work just as well with $\phi_1 > \phi_2 > \phi_3 > \ldots > \phi_L$. Furthermore, Cohen showed that whether or not buffer capacity exists, is irrelevant to the solution.

The problem is solved as a nonlinear programming problem, where:
• The Objective Function is to minimize the makespan for the production of M products.
• The variables are:

One) t_1 - time for the first cycle.

Two) Q_{ij} - intersecting point between Stations i and j. Each intersecting point is associated with two segments of adjacent learning curves. Therefore, there are "2L-1" variables.

Three) The constraints
  (a) The sum of station Standard Time equals the Standard Time for the whole job
  (b) At each intersecting point, the performance times for the intersecting learning curve are identical.
  (c) The intersecting points are ordered according to the ascending learning slopes of the various stations.

An Illustration

Consider a 3-station assembly line required to produce 100 products whose overall standard time is 20.0 hrs. Learning slopes from Stations A, B and C are 70%, 80% and 90%, respectively.

Using the GINO system, the problem was solved with the following results:

t_{A1} - 81.61 hrs.

t_{B1} - 50.38 hrs.

t_{C1} - 26.9 hrs.

Intersecting point Q_1 occurs at n = 10 cycles.

Intersecting point Q_2 occurs at n = 30 cycles.

Makespan (for 100 products) = 1957 hours.

Makespan for equal work allocation = 2302 hours (a 17.6% increase!!)

The first cycle times for the three stations are in descending order, according to their learning slopes, but the standard time allocations are not necessarily in descending order as seen in Table 6.2.

Table 6.2: Optimizing the Makespan for $\phi_A < \phi_B < \phi_C$

Station #	ϕ	$\left(t_1/C\right)$	t_1	Station Allocation
A	70	15	81.6	5.45
B	80	9	50.3	5.59
C	90	3	26.9	8.96
				Σ 20.00

The main problem with this model is technical. No matter what the allocation to each station is going to be, their respective learning slopes remain unaltered and we realize this is unlikely. However, heuristic solutions can be found. In his thesis, Cohen (1992) covers many other problems of a similar nature, but this is outside the scope of this book. While the analysis of adjacent stations having different learning slopes is interesting, it must be understood that optimal balance conditions are obtained when the learning slopes between adjacent stations are equal (as far as possible). Goralnick (1998) showed that stations of equal balance, whose learning slopes were approximately equal, gave the highest performances. Variations in the order of about $\pm 2\%$ in learning slopes of adjacent stations still resulted in high performances.

Goralnick's work shows that both balance and station learning slopes need to be balanced in order to obtain the best results for assembly line performance.

6.2.3 LC with mixed-model assembly lines. Mixed-model assembly lines are used when several models of the same product are assembled on a common assembly line. Mixed-model assembly lines are commonly used in industry (see, for example, Schonberger, 1982; Monden, 1983. Chakravarty and Shtub (1985, 1986, 1992) address the effect of learning on the operation of mixed-model assembly lines. Part of their work was based on an earlier paper by Thomopoulos and Lehman (1969), who developed a learning model for the mixed-model assembly line. However, Chakravarty and Shtub go much further; their objective was to minimize inventory-carrying costs, set-up costs and labor costs, subject to dynamic capacity constraints dictated by the learning effect. To do this (among other steps), required an analysis of:
(a) the accumulation of learning over time, with task times reducing according to the power curve, and
(b) simultaneous learning on several models in a model-mix environment.

Their development of the solution brings into play aspects of scheduling, inventory and set-up costs, and task precedence, which leads to a comprehensive method for analyzing the mixed-model assembly line with learning. In effect, the solution provides a complete solution to the mixed-model sequencing problem, which this writer believes is outside the scope of this book.

Interested readers should read both references given above in order to get a balanced prospective on this topic.

6.2.4 LC in assembly lines for new products. A model for using LC theory in assembly lines for new products was presented by Dar-El and Rubinovitz (1991), who described the application for the assembly of aircraft engines built by the Israel Aircraft Industries Ltd. The engine housings are made of composite materials with which the IAI plant had limited experience.   Consequently, initial task time estimates were inaccurate, and required to be updated using actual execution times that were generated. Clearly, learning was certain to take place and would need to be accounted for in order to plan efficient line balances for the entire production run. The task times were gradually reduced as more units were produced, but the learning effect was not uniform for all tasks. Stations on the line were allocated different tasks, so that different learning rates could be expected at each station for production of subsequent units. As a result, a production line balanced for the first unit of product will soon be out of balance, as learning takes place.   Rebalancing the line for subsequent production units involves the risk of potential learning losses for tasks that are reassigned to new stations.

The jigs used for production are very expensive and would not be duplicated. This means that assembly operations requiring the same jig would be constrained to the same single station. Also, due to inexperience with composite materials and complexity of the manufacturing process, the arrival time of parts needed for the assembly could not be guaranteed, so that updating of part due dates had to be considered by the planning model.

In this paper, we consider rebalancing of the assembly line to compensate for imbalance introduced due to learning at subsequent products. Such rebalancing must be performed so as to minimize learning losses due to task assignment to new stations. Additional constraints, due to zoning and restrictions imposed by expensive jigs and tooling, were also considered by the balancing algorithm. To provide for these realistic and severe constraints and still achieve good line balance, the powerful MUST algorithm was used (Dar-El and Rubinovitz, 1979). This algorithm uses a frontier search method, combined with efficient computer techniques, to generate all optimal solutions for a given line-balancing problem.

The procedure uses the methods of Chapter 4 to find values for "b" and "t_1" for each station. However, on-line data on station performance are fed into the computer at the start and finish of work at each station. A method was developed that determines when it is economically worthwhile to drop a station from the line since the production rate should not be raised and learning would have greatly increased the productive capacity of the system.

When a station is dropped, the work performed in this station must be redistributed to the remaining stations in the line; the balancing algorithm ensures that most of the original work done at these stations remain at the stations when rebalancing takes place, so as to retain the attained learning advantages.

The model was successfully implemented for on-line planning and scheduling of a new production line for the assembly of aircraft engine housings by the Israel Aircraft Industries Ltd.

The initial quantity of assembled units was small - about 50 units were ordered. Assembly labor time accounted for about 30% of total cost. This fact was of major concern to project management, since any inefficiencies or idle times during assembly could contribute significantly to the profitability (or loss) of the entire project.

The implementation of the model and initial planning of the engine-housing assembly line was performed as a graduating-year student project in Industrial Engineering. One of the students involved in the project is now employed by IAI. The company adopted the model and is using it for continuous on-line scheduling and rebalancing, while updating the learning parameters as a result of accumulated actual performance times for each unit.

6.2.5 Optimizing "cycle time: number of stations" under learning constraints.
Assembly lines can be considered as a special case of production lines in which the manufacturing capacity utilized (i.e., the number of operators) can be varied to meet the capacity needs of the demand. When output just equals the demand, a typical 'EOQ' formula would indicate an indefinitely large economic order quantity - meaning, that the assembly line should operate continuously and the inventory of finished products would be zero.

However, this traditional approach ignores an important cost factor that helped to justify the use of assembly lines in the first place, namely, the cost of 'learning'. We know that it takes a lot less time to 'learn' a simple task (i.e., in order to achieve standard performance), than it takes to learn longer, complicated tasks. Low demands for products over extended periods usually require much longer cycle times, which in turn, require longer learning periods (Rosenwasser, 1982). As the required cycle time increases, the learning costs also increase. This suggests that conditions may arise for accumulating inventory of finished products to supply demand during periods of non-production. Batch production also has the advantage of being better able to cope with changes in demand, since, with continuous production, the only realistic alternative is to vary the number of hours worked per day. However, batch production also means an increased number of setups and possibly, some changes in the facilities needed.

The assembly conditions tackled in this paper assume a given continuous demand for a single product taken over a specific planning horizon. The objective is to find the EOQ and 'optimal cycle time, station number' combination (henceforth, referred to as the 'C_0^*, L^*' combination) that minimizes the overall assembly costs based on the following five factors: learning costs, inventory costs, setup costs, balancing costs and facility costs.

The learning cost was developed on an 'older' assumption that the total number of cycles (G) to reach the standard time was a linear function of the station standard time, C, itself (this work was begun about 1979). Thus,

$$C = t_1 \cdot (G)^{-b}$$

where

$$t_1 = C \cdot (G)^{-b}$$

and

$$G = a + d \cdot C$$

with the range of values for a & d empirically found (Rabinovitch, 1983) as,

$$a = 300\text{–}500 \; ; \; d = 300\text{–}700.$$

Today, we now know this to be an incorrect assumption, with "b" and "t_1" being predicted according to Section 4.2 and G calculated on the basis of Section 4.4.

Anyway, the approach was the first attempt at balancing learning costs against all other costs, such as, inventory, setup, facility and balancing costs (see Dar-El and Rabinovitch, 1988).

Figure 6.5[*] is taken from Dar-El and Rabinovitch's article which shows RELC (the ratio of the optimized cycle time to the cycle time for continuous production) versus the production order, for varying values of the station learning slopes.

*Use of copyright material by *Int. J. Production Research* is gratefully acknowledged.

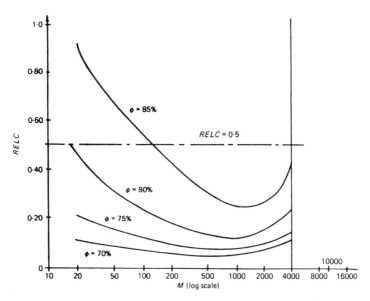

Figure 6.5: RELC versus total demand M, for various learning slopes

Maximum savings are obtained for a fairly wide range of RELC values, especially for the 70% and 75% learning slopes, with production quantities in the order of 1000 items.

This work really needs to be redone using a more correct basis for determining b, t_1 and G, but the paper does illustrate the application of learning curve theory in finding the optimal "C, L" combination.

6.3 LC with Long Cycle Time Tasks

Two papers have been written on this topic, one by Globerson and Shtub (1984) and the other by Dar-El, Ayas and Gilad (1995). Because of its ramifications to the growing high-tech and military industries, this topic is becoming increasingly important since, task cycle times are very large (counted in days and weeks) while order quantities are small (most are less than 50 items).

Actually, the early work by Globerson and Shtub (1984) has been superceded in virtually every aspect by Cohen's (1992) work covered earlier in Section 6.2.2. Besides, their starting assumptions are without justification. Referring to "N" repetitions needed to achieve the standard time (such as from MTM), they say, "Estimates of the values of "N" and "m" (the learning constant) may be obtained from company files". This has no justification, since there is no published data anywhere on values of N for particular tasks (as there are tables for "b" (or "m" used in their paper)). However, this was an early paper, while knowledge on the subject was just beginning to accumulate.

The second paper, by Dar-El et al. (1995), has a much firmer scientific basis to their work on estimating the power curve from learning parameters for long cycle time tasks. The basis of their work led to the prediction of "b" and "t_i" discussed in Chapter 4.

6.3.1 A model for learning behavior in long cycle time tasks[**]. Chapter 5, Section 5.2.5 discusses the aspect of forgetting in long cycle time tasks, which is assumed to consist of a series of non-repetitive unique sub-tasks, whose standard times average, about 1½ minutes, as illustrated in Figure 5.7. The model assumes independence of the sub-tasks and there is no transfer of skills between them.

The assembly of a typical pop-up toaster, for example, may comprise four sub-tasks: the base assembly, the electrical components, the mechanical components, and the body assembly. In fact, this may be the very manner in which a four-station assembly line for this pop-up toaster would be built. The same applies to the many assembly lines found in plants producing domestic appliances and industrial products.

Once a sub-task (or simply 'task') is completed, it is assumed to appear again only in its correct sequence on the next cycle. Thus, the repetition of the task is interrupted by a time approximated by the cycle time itself. This interruption causes some 'forgetting' to occur; the extent of forgetting is clearly a function of the interruption length, or, the cycle time itself.

We now develop expressions for calculating the learning behavior of long cycle tasks.

Let w = the number of short cycle tasks in a long cycle task,

   $t_i^{(1)}$ = the execution time for the first cycle of short task i,

        $i = 1,2,\ldots w.$

   b_i = the learning constant for short task i, with no forgetting.

Then, with no forgetting, $t_i^{(n)}$, the execution time for the n-th repetition on task i, is given by,

$$t_i^{(n)} = t_i^{(1)} \cdot n^{-b_i} \quad .$$ (6.7)

[**] Use of copyright material by *IIE Transactions* is gratefully acknowledged.

Now, according to our classification method, b_i can only have one of four values, each associated with one of category C1, C2, M2 or M1. Call these b_{c1}, b_{c2}, b_{m2} and b_{m1}, respectively.

Let $\hat{t}_{c1}^{(1)}$, $\hat{t}_{c2}^{(1)}$, $\hat{t}_{m2}^{(1)}$ and $\hat{t}_{m1}^{(1)}$ represent the sum of all first cycle times whose learning constants are b_{c1}, b_{c2}, b_{m2} and b_{m1}, respectively. Then T_1, the first cycle time for the overall task, is given by

$$T_1 = \hat{t}_{c1}^{(1)} + \hat{t}_{c2}^{(1)} + \hat{t}_{m2}^{(1)} + \hat{t}_{m1}^{(1)}. \tag{6.8}$$

The execution time for the n-th cycle of the long cycle task, *without forgetting*, is therefore,

$$T_n = \hat{t}_{c1}^{(1)} \cdot n^{-b_{c1}} + \hat{t}_{c2}^{(1)} \cdot n^{-b_{c2}} + \hat{t}_{m2}^{(1)} \cdot n^{-b_{M2}} + \hat{t}_{m1}^{(1)} \cdot n^{-b_{M1}}. \tag{6.9}$$

However, $\hat{t}_{c1}^{(1)}$ can be expressed as

$$\hat{t}_{c1}^{(1)} = (t_1/C)_{c1} \cdot \hat{C}_{c1} = S_{c1} \cdot \hat{C}_{c1}, \tag{6.10a}$$

where $\hat{C}_{c1}$ is the sum of the standard times of all tasks in category C1, and $S_{c1} = (t_1/C)_{c1}$, i.e., the ratio of the execution time for the first cycle divided by the standard time for elements in category C1.
    Similarly,

$$\hat{t}_{c2}^{(1)} = S_{M2} \cdot \hat{C}_{C2}, \tag{6.10b}$$

$$\hat{t}_{m2}^{(1)} = S_{M2} \cdot \hat{C}_{M2}, \tag{6.10c}$$

and

$$\hat{t}_{m1}^{(1)} = S_{M1} \cdot \hat{C}_{M1} \tag{6.10d}$$

We can use (6.9) and (6.10) to express TT*(n)*, the total time for completing *n* products with *no forgetting*, i.e.,

$$TT(n) \approx \frac{S_{c1} \cdot \hat{C}_{c1} \cdot n^{1-b_{c1}}}{1-b_{c1}} + \frac{S_{c2} \cdot \hat{C}_{c2} \cdot n^{1-b_{c2}}}{1-b_{c2}} +$$

$$\frac{S_{m2} \cdot \hat{C}_{m2} \cdot n^{1-b_{m2}}}{1-b_{m2}} + \frac{S_{m1} \cdot \hat{C}_{m1} \cdot n^{1-b_{n1}}}{1-b_{m1}}. \qquad (6.11)$$

When forgetting occurs, the times for the first cycle are unaffected. Therefore, (6.8) and (6.10(a-d)) still apply. However, learning rates are decreased, with b_{c1} becoming b_{c1}^{f}, the learning constant *with* forgetting $(b_{c1}^{f} < b_{c1})$; b_{c2} becomes b_{c2}^{f}, b_{m2} becomes b_{m2}^{f} and b_{m1} becomes b_{m1}^{f}.

Then, $TT(n)^{f}$, the total time *with* forgetting becomes

$$TT(n)^{f} \approx \frac{S_{c1} \cdot \hat{C}_{c1} \cdot n^{1-b_{c1}^{f}}}{1-b_{c1}^{f}} + \frac{S_{c2} \cdot \hat{C}_{c2} \cdot n^{1-b_{c2}^{f}}}{1-b_{c2}^{f}} +$$

$$\frac{S_{m2} \cdot \hat{C}_{m2} \cdot n^{1-b_{m2}^{f}}}{1-b_{m2}^{f}} + \frac{S_{m1} \cdot \hat{C}_{m1} \cdot n^{1-b_{m1}^{f}}}{1-b_{m1}^{f}} \qquad (6.12)$$

Looking at (6.12), we see that the $\hat{C}_{j}$ values are estimated as the task standard times; the S_{j} values are obtained from Figure 4.2 leaving only the b_{j}^{f} values to be determined.

6.3.2 Estimating the learning constant "b" under conditions of forgetting. Referring once again to Section 5.2.5, which described Sabag's (1988) experiments with a task requiring three sub-tasks, A, B and C. The next experiment was designed to understand the effects of forgetting. Four subjects were used in each experimental condition and no subject participated in more than one experiment. Table 6.3 describes the experimental conditions, it being emphasized that each subject would perform the task *just once*, then do it again *just once* after the appropriate interruption had passed. This way, we would simulate the conditions described by Figure 5.4, for products where the cycle time varied from one to seven days.

Table 6.3: Experimental Setup for Estimating Learning Constant under Conditions of Forgetting

Interruption time between complete assemblies (days)	Twenty complete assemblies using:	
	Method 1 (ordered) Learning Group C2	Method 1 (random) Learning Group M2
1	X	X
2	X	X
7	X	X

The implied assumption of task independence is readily maintained, because the subjects, all university students, spent the interruption period on study and leisure activities, with no possibility of 'skill transfer' occurring. In actual industrial settings, a strong argument may be made for some transfer of skills to occur between several of the short cycle tasks in a long cycle time task. Experimental results are given in Table 6.4.

Table 6.4: Learning Constants Found in the Experiments

Interruption period (days)	b^f, the learning constant using:	
	Method 1 (ordered) Learning Group C2	Method 1 (random) Learning Group M2
1	0.385	0.161
2	0.348	NA
7	0.317	0.137

Figure 6.6 illustrates the relationship between b^f and b_0, the learning constant with forgetting and without forgetting, respectively. The graphs tend to follow a power decay form, with the relationship empirically determined as

$$b^f = b_0 \cdot (q+1)^{-0.152},\qquad(6.13)$$

where q is the interruption period in 'days'.

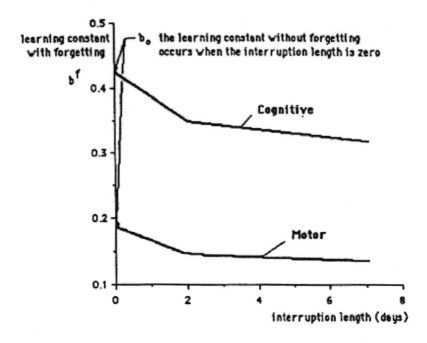

Figure 6.6: Influence of interruption length on the learning constant

Table 6.5 compares the experimental results with appropriate values determined from (6.13).

Obviously, there are insufficient data to fully support (6.13), though the applications to both C2 and M2 data are fairly good. Until more data are available, the learning constant with forgetting can be determined from (6.13).

Table 6.5: Learning Constants with Forgetting Experimental and Predicted

Interruption period (days)	High Cognitive (C2)		High Motor (M2)	
	Exp'tal	(Eqn (6.13))	Exp'tal	(Eqn (6.13))
0	0.423	(0.423)	0.186	(0.186)
1	0.385	(0.381)	0.161	(0.167)
2	0.348	(0.358)	NA	(0.157)
7	0.317	(0.308)	0.137	(0.136)

We are now able to predict the execution times of long cycle time tasks.

6.3.3 Predicting performance times for long cycle times. The method proposed for determining performance times is as follows:

1. Divide the overall task into its constituent sub-tasks.
2. Estimate the standard time, C_i, for each sub-task i. (Use MTM, Work Factor, MODAPTS, etc. or standard data, if available).
3. Categorize the learning slope of sub-task i, as C1, C2, M2 or M1 (Section 2.1, A).
4. Group each sub-task into one of the four categories, k, where k = C1, C2, M2 or M1. Within each group k, the ratio $S_k = (t_1/C)_k$ is constant; so are the b_j^f values, $\forall i \in k$. Add the standard times of all sub-tasks in each group, i.e., $\hat{C}_k = \sum_{i \in k} C_i, k = C1, C2, M2$ and M1.
5. Estimate $t_i^{(1)}$, the time for the first cycle of sub-task, i, from its (t(1)/C) value (see Figure 4.2) multiplied by C_i.
6. Use (6.13), which accounts for forgetting, to revise the learning constant estimate, i.e., obtain b_j^f, $\forall$ sub-tasks i.
7. Evaluate (6.12) to obtain the total time for completing the required production.

Table 6.6 compares the predicted assembly time for 20 products with the average time taken by seven subjects. The difference of 4.4% must be considered as being highly satisfactory.

Table 6.6: Comparing Actual with Estimated Times for 20 Assemblies Using Method 1

Actual			Estimated (eq'n (6.12))			Absolute value of
t_1	b	TT(20) (min.)	S_k	$\hat{C}_k$	TT(20) (min.)	error (%)
29.56	0.423	288.6	12.8	2.53	301.3	4.4%

6.4

6.5 LC in Batch Size, or Lot Size Production

It frequently occurs that a large order for items to be supplied over a year, or more, are produced in batches throughout the year. When the rate of demand is constant, we get the well known 'saw-tooth' diagram for inventory levels and the EOQ (Economic Order Quantity) analysis appears in every book published on Production, or Operations Management.

However, not that many books include the conditions of learning in their analysis. Two production strategies are covered: one, when a fixed quantity (lot or batch size) production is periodically executed, or secondly, periodically, using equal production time periods. The former should be the more appropriate strategy to use since it provides for easier quantity planning and control.

Production over the planning horizon is illustrated in Figure 6.7, which shows the effects of learning during the early production runs.

Inventory is accumulated during the production run period $v(i)$ at the rate of $\{\bar{p}(i) - r\}$, where $\bar{p}(i)$ is the average production rate during $v(i)$ and r is the constant consumption rate. In reality, the production rate continuously changes with respect to time, but these are generally only meaningful for the first batch (or two at the most), when learning is rapid. Nander and Adler (1974), important contributors in this topic, also assume that production rates are constant over the appropriate production periods.

The rest is pure mathematical manipulation of functions relating to inventory costs. All researchers on this topic have simply assumed that the power curve applies and estimates of 'b' and 't_1' are 'magically' available – but then, Learning Curve Theory is not central to their work, as in Inventory Theory.

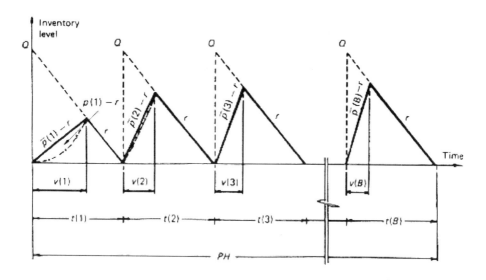

Figure 6.7: Production schedule and inventory level graph

From a historical perspective, "pioneering" research work on the incorporation of learning and forgetting into the determination of economic lot sizes was begun by Keachie and Fontana (1966). Forgetting can be regarded as retrogression in learning, which causes a loss of labor efficiency due to breaks between batch production runs. Steedman (1970) and Muth and Spremann (1983) extended the work of Keachie and Fontana to more general optimal lot-size models. Researchers in this interesting field included Spradlin and Pierce, 1967; Adler and Nanda, 1974a, 1974b; Sule, 1978, 1981, 1983; Fisk and Ballou, 1982; Kopcso and Nemitz, 1983; Wortham and Mayyasi, 1972; Smunt and Morton, 1985; Klastorin and Moinzadeh, 1989; and Nanda and Nam, 1992, 1993. All the above works assume continuous demand, but Donath et al. (1981) and Chiu (1997) have considered discrete time-varying demands.

All researchers on this topic have assumed that learning follows the power curve. However, with forgetting, the assumptions vary. Some say it should be ignored, since relearning is very rapid, while most researchers conjure up some function to represent the forgetting effect. None of these authors have any laboratory or empirical evidence to support their assumptions, though reference was made to research work dealing with forgetting (e.g., Globerson et al., 1989). These works were done, however, when "forgetting theory" was at its infancy, meaning that they do not provide much scientific support for learning-forgetting in the batch-size or lot-size models.

6.6 LC in a Group Technology (GT) Environment

Job-shop work (where quantities are very small, often just one) and batch production, together constitute the large bulk of manufacturing operations today. Traditionally, plant layouts were based on process requirements, so that each department in a plant would segregate similar processes (e.g., the press shop, all drilling machines, the turning shop, etc.). The result was that many products were required to travel excessive distances in order to be processed, especially if a particular type of process was needed more than once. Years ago, this led to the development of Group Technology (GT), in which parts of products are identified and grouped together to take advantage of their similarities (Eckert, 1984). Each family would possess similar manufacturing characteristics, yielding a reduction in setup time, lowering in-process inventories, simplifying scheduling, improving tool control and extending the use of standardized process plans.

The development and use of standard process plans in manufacturing has learning implications, especially when learning is associated with the process by which a product is made, rather than the product itself. The intensity of the impact depends on the degree of standardization in the process. Reduced setup times can be obtained by sequentially ordering parts which require the same or very similar processing at a particular production stage. If these different parts require identical processing, there is no doubt that some shared process experiences will result.

The end result is that most job shops and batch production shops have been transferred from process flow layouts to "mini-product" flow layouts, taking the form of GT cells where inventory levels and lead times are sharply reduced and production flow increased. The implications for learning are clear, since the processes in a GT cell are virtually identical for a large number of different parts, then learning must ensue with processing time reduced with experience. Investigations show Globerson and Millen (1989) to be the first to indicate the relationship between GT production and learning, producing some simple models that show the potential gain by considering the learning effects. But, strangely, this aspect has not been picked up by other GT researchers in the last decade. Instead, they have concentrated on the technology of GT grouping and other factors (such as scheduling).

Learning Curve application to GT remains a promising area for future research.

6.7 Learning in a JIT Environment

The JIT concept was first popularized by Monden (1983) about 20 years ago. The production control system included in the JIT methodology is the KANBAN system, which can be described as "Pull-Closed"; "Pull" means that production is initiated at a subsequent process by pulling in the raw materials from the previous process, and "Closed" means that the total number of KANBAN cards used in the system is limited (or, fixed).

It took about a decade before the first research on KANBAN systems began appearing in the literature (see Person, 1989). The KANBAN system provides the production control mechanism for both production lines, as well as assembly lines, and over the last few years, many industrial assembly lines have been designed with KANBAN control (this is a 'hot' topic with large manufacturers as well as consultants).

However, none of the research nor applications have incorporated learning effects in their design, that is, not until the publication by Swensett, Muralidhar and Wilson (1993). These authors carry out several simulation studies on the behavior of production lines under KANBAN control. They assume that learning takes place (the model employed is not given, but it appears to be the power model that was used) and the simulation was used in order to determine the number of KANBAN cards needed in order to meet a desired level of production (at least 90%). Naturally, the production line capacity increases substantially, but there is very interesting data on the behavior of the number of KANBAN cards (i.e., the in-process inventory) needed with respect to the total quantity produced.

As an application area, it is only a simulation study with computer generated data and merely shows the potential benefits of including learning into JIT lines. We await the results, of case studies taken under industrial conditions, to learn more of the problem and likely benefits accruing in such applications.

6.8 'Speed – Accuracy' in a Learning Environment

Psychological research has indicated that individuals who tend to be faster, are usually more accurate, especially if it is associated with a greater instruction emphasis (see Pachella, 1974; Pew, 1969; Fitts, 1966; Wickens, 1984). It was further shown by Howell and Kreidler (1963) that feedback which does not conform to instructions creates learning interference.

From the engineering side, only a very few references emerge that deal with speed-accuracy aspects (see Buck and Cheng, 1993; Vollichman, 1993; and Erdal, 1998). Buck and Cheng (1993) expanded their study on speed-accuracy to include the fit to different Learning Curve models (the power model and exponential model). Vollichman (1993) did his M.Sc. thesis on speed-accuracy under learning-forgetting-relearning conditions. Erdal (1998) applied the speed-accuracy studies on an electronic counter assembly and came to the early conclusion that with adequate training, errors simply did not appear! Erdal's was a simple procedural task, whereas the other two had high cognitive and decision-making elements built into the task.

Discussions on this topic will cover the two papers:
    6.7.1 Speed-Accuracy via Buck and Cheng (1993)
    6.7.2 Speed-Accuracy via Vollichman (1993).

6.7.1 Speed-Accuracy via Buck and Cheng (1993)[*]

The Equipment
"A task simulator consisted of a metallic box with a linear array of six lights and six adjacent push-button switches. These switches changed state at each actuation but they physically returned to the original position after an actuation. Thus, *the subject could not tell which switches had been actuated during the trials by a visual observation of the switches*. The actuation sequence had to be remembered. Connections of the switches to the lights were randomized within the metallic box so that the sequence of lights and switches differed and trial-and-error practice was needed to discover the correspondence between the switches and lights. Stopwatches were used to time the subjects, and those time values were recorded. The number of switch actuations were also directly observed and recorded. Electronic records were made on the timing and actuation of switches for verification."

The Procedure
"Each subject was individually seated at a table with the task simulator on the table directly in front, always in the same orientation. All of the lights on the device were off. The subject was instructed to actuate the switches so that the lights in the array would turn on in their left-to-right sequence. If any activated or deactivated light was turned on out of sequence, then all of the out-of-sequence lights must be first turned off before proceeding further. Note that turning out a light which was accidentally turned on required a second switch actuation. The time it took the subject to turn on all six lights was timed by the experimenters and the

---

[*] Use of copyright material by *IIE Transactions* is gratefully acknowledged.

number of errors made in switch actuation was recorded in each trial. After each trial, feedback was given to the subject. Ten consecutive trials were administered to each subject."

Other instructions and feedback differed among the three groups of subjects. One group, known as the *speed group*, was instructed to perform the tasks as quickly as possible without regards to mistakes. Feedback of results was given after each trial on the actual time spent but nothing was said to the subjects of this group about switch actuation errors. In fact, the experimenters concealed the collection of error data from the subjects of this group. A second group, known as the *accuracy group*, was instructed to work as accurately as possible without regard to time, and they were given feedback on the number of errors made after each trial. The collection of performance time data was concealed. The third group, *speed-accuracy group*, was instructed that speed and accuracy were equally important, and they were given both the time and errors as feedback at the end of each trial.

Only results that are relevant to our topic are included in Table 6.7.

Our important finding is that the total errors associated under 'Speed' *are less* than those of 'Accuracy' and the 'Speed/Accuracy' combination. This also fits in with the earlier remarks made on this topic.

Table 6.7: Summary of Relevant Data (Buck and Cheng, 1993)

Criterion	Accuracy		Speed/Accuracy		Speed	
	Power	Expo	Power	Expo.	Power	Expo.
Average Total Errors	4.24	2.92	3.07	2.76	3.02	2.82
Cum. Av. Total Errors	7.68	1.43	5.61	1.23	4.61	1.04

6.7.2   Speed-Accuracy via Vollichman (1993).   Visits to several processing plants were made prior to the start of the experiments with attention focussed on control room operations. The purpose was to observe the actions taken by control personnel in reaction to various disturbances occurring to the operating system. Processing plants were chosen for study because of the availability of a simulation model of a processing plant in the Safety Center of the Technion, which we intended to use for our experiments.

The question of "accuracy" within the context of a processing plant is easily defined as the types of errors that occur during operations. However, "accuracy" is also an outcome of the "effort invested" in obtaining error-free performances. "Effort invested" is translated as the "level of quality requirements" demanded of the operators during the performance of their tasks.

Three quality groups (H, M, N) were defined for operating the assignment. The first group, H, called the *high quality requirements group* was instructed to operate without making any mistakes (if at all possible). The people in Group H were told that while they were likely to make mistakes at the beginning of the experiment, they had to concentrate on diminishing errors. They were also told to improve their performances but maintaining the quality of operations was by far their main objective. Group M was called the *middle quality requirements group*, who were permitted to make mistakes of Types A and B (minor errors), but as far as possible were not to make mistakes of Types C and D (major errors). Group N was used as a *control group with no specific quality instructions*; subjects were asked to improve performance times without mentioning any quality requirements. While being aware of malfunctioning and operational errors, it was emphasized that *their* main goal was to improve their performance times even if this came at the expense of performance quality. Nevertheless, they should still try to operate, as far as possible, without causing errors.

The following research objectives were defined:

(One) To investigate learning behavior under specific quality requirements;

(Two) To investigate the experience-accuracy (quality) characteristics;

(Three) To investigate the speed-accuracy (quality) characteristics;

(Four) To investigate accuracy (quality) characteristics under conditions of "forgetting".

The Subjects

A total of 48 subjects were used for the experiments and were selected with the following data:

(One) 34 men ages 20-28 and 14 women ages 20-27.

(Two) All subjects had at least 12 years of schooling, with 18 having a BA or B.Sc. degree.

(Three) Most subjects were experienced in working with computers.

(Four) None of the subjects had any experience in working with the specific simulator, nor any other similar computer simulation program.

(Five) Participation of the subjects in the experiment was on a voluntary basis; they were not paid to participate in the experiment, nor did they get any financial incentive to motivate their performances.

Starting Conditions

Subjects were put through the following steps before the experiments began:

(One) Reading a general explanation about the system and its operation during normal operating conditions.

(Two) Receiving a verbal explanation about the system including a demonstration.

(Three) Exercising operation of the system under normal conditions for 20 minutes.

(Four) Reading a general explanation of the system during time of emergency, the type of malfunction of the filling system and the ways to deal with malfunctions.

(Five) Demonstrating the operation of the malfunctions during times of emergency.

(Six) Getting an explanation on the specific experiment, the expected treatment routine during malfunctioning and the order of the experiments.

Experiment 1

(a)   The subjects were divided into three groups:

     Group H   16 subjects under high quality requirements
     Group M   16 subjects under middle quality requirements
     Group N   16 subjects with no quality requirements.

(b)   Each subject performed 16 repetitions of the assignment.

(c)   Two operation indices were measured for each experiment.
     (One) Time of performance;
     (Two) Number of mistakes from each of the four types.

Experiment 2

(a)   Four values of stoppage duration were defined: a week, two weeks, a month and two months.

(b)   Each group of subjects from Experiment 1 (16 subjects in all) was divided into 4 subgroups according to the four stoppage durations.

(c)   In this way, there were 16 different experimental groups with 4 subjects in each. An experimental group's work is characterized by the required level of quality and the stoppage duration.

(d)   Considering the time constraints, it was decided that the number of "learning-forgetting-relearning (L-F-R)" cycles for the experiment would be 4, 4, 3, 2 according to the four groups of stoppage durations.
(e)   In each cycle, each subject performs 16 repetitions of the assignment.
(f)   Two operation indices were measured:
        (One) Time of performance;
        (Two) Number of mistakes from each of the four types.

Results of Experiment 1 is best shown in Table 6.8 and Figure 6.8 which illustrates the total number of errors associated with each repetition. Note that all three quality conditions eventually result in zero errors by the completion of about 15 repetitions.

Table 6.8: Data for Experiment 1 (Vollichman, 1993)

Group	t_1	B	ϕ	R^2
H (High)	162	0.261	83.5	0.71
M (Medium)	156	0.371	77.4	0.71
N (No Quality Specified)	141	0.464	72.5	0.79

Results from Experiment 2 are just as dramatic. Figure 6.9 shows the learning-forgetting-relearning for Group H only under varying interruption breaks (of up to 2 months) and shows the total errors occurring. Again, as the cycles progress, the number and frequency of errors disappear altogether.

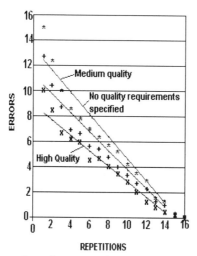

Figure 6.8: Total number of errors under varying quality requirements

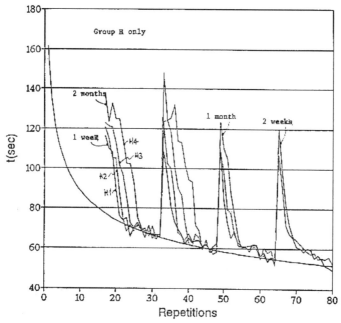

Figure 6.9: Learning-Forgetting-Relearning for Group H (High quality requirements)

Table 6.9 shows the number of repetitions needed to return to the original learning curve. We see these numbers getting progressively smaller as the number of cycles increase. Furthermore, the original learning curve is reached very rapidly.

Table 6.9: Maximum Number of Repetitions to Return to the Original Learning Curve

"Stoppage relearning" cycle number	No Quality Requirements			
	week	2 weeks	month	2 months
1st break	4	4	5	7
2nd break	2	3	4	6
3rd break	2	2	3	
4th break	1	2		

"Stoppage relearning" cycle number	Medium Quality Requirements			
	week	2 weeks	month	2 months
1st break	4	5	6	7
2nd break	3	3	6	7
3rd break	3	3	5	
4th break	1	3		

"Stoppage relearning" cycle number	High Quality Requirements			
	week	2 weeks	month	2 months
1st break	4	5	7	9
2nd break	3	4	6	9
3rd break	3	4	5	
4th break	2	3		

Finally, Table 6.10 shows the maximum number of repetitions to achieve error-free performance and this is true also for the 2-month interruptions.

Table 6.10: Maximum Number of Repetitions to Error-Free Performance

No.	High Quality Requirements			
	week	2 weeks	month	2 months
1	3	3	4	5
2	2	2	4	5
3	1	2	3	

4	0	1		

No.	Medium Quality Requirements			
	week	2 weeks	month	2 months
1	3	4	4	5
2	3	4	4	5
3	2	2	3	
4	1	2		

No.	No Quality Requirements			
	week	2 weeks	month	2 months
1	4	5	6	6
2	4	5	5	6
3	2	3	3	
4	2	2		

6.7.3 Discussion. The most significant results arising from this work is that very few repetitions are needed to return to error-free performance, even after long stoppages. Furthermore, these recovery repetitions are reduced even further with experience. Also significant is that error-free performances occur with *all* three types of quality performances, but that the group asked to concentrate on 'speed' was achieving error-free performance about 2½ times *faster* than group H that was given the strongest quality requirements.

The implication for training schedules to counteract hazardous safety conditions becomes very clear. If it were somehow possible to simulate the hazardous conditions and train operators in the appropriate countermeasures, then we could begin to determine efficient and effective training (or, retraining) schedules that can result in very rapid, error-free countermeasures being taken, which will either avoid the 'accident' or else, reduce its potential damage.

6.9 Learning Curves in Machining Operations

This section summarizes some of the pioneering work done by Nadler and Smith (1963) in Machine Shops. Table 4 (in Chapter 4) was, in fact, a summary of their findings for learning slope values determined for common machining processes.

As expected, they concluded that each basic operation had its unique learning slope and the same operation may have different values in different companies.

Nadler and Smith also used a simple formula developed by McCambell and McQueen (1956) to estimate the learning slope for a product composed of several components. Thus,

$$C = X_1 Y_1 + X_2 Y_2 + \cdots + X_n Y_n \qquad (6.14)$$

where,
C   is the learning slope for the product
X_i is the percentage of labor needed to produce part i
Y_i is the learning slope for producing part i
n   is the total number of parts comprising the product.

Equation (6.14) can also be used for calculating the learning slope of a complex part, so that Y_i would be the learning slope for each separate operation and X_i the percentage of labor needed to complete that operation.

Nadler and Smith applied equation (6.14) to eight products whose pertinent results are given in Table 6.11.

Two conclusions can be drawn from their examples:
(a) The error in slope is quite small.
(b) Their estimates for eight product learning slopes involving Machine Shop operations had a fairly narrow range: from 80.7% to 87.4% (leaving out the 77.5% case, the measured range would be 82.3% to 87.4%).

Table 6.11: Summary of Product Learning (from Nadler and Smith, 1963)

Product	Learning Slopes		Differences	Absolute
	Calculated	Actual	(Theo-Act)	Error-Free (%)
A	82.8	83.0	-0.22	0.27
B	82.6	82.3	0.32	0.39
C	82.4	77.5	4.90	5.95
D	84.8	84.6	0.19	0.22
E	85.6	87.4	1.78	2.08
F	83.3	85.6	2.39	2.87
G	80.7	84.6	3.82	4.74
H	83.3	84.4	1.12	1.35
Average Error of Estimate = 2.22%				

Can this suggest that product learning slopes for a specific company hovers around some central value without too large a variance?

7 COST MODELS FOR OPTIMAL TRAINING SCHEDULES

Regular training schedules are used for individuals who work on tasks whose failure could cause loss of life and/or considerable material damage. We consequently require such individuals to undergo regular refresher training courses in order to get their skills back to peak performance and thereby reduce the chances of making a response error to dangerous situations.

Typical examples where such training regularly occurs include: Fighter pilots exposed to bombing missions, or, offence/defense tactics against enemy aircraft; Passenger aircraft crews (pilot, co-pilot and navigators); Flight controllers is another well known application used in both civilian and military sectors; There are many applications in the military area – virtually every control room contains such requirements, especially those who work on counteracting incoming missiles, jamming radio signals, and so on. Taken less seriously, but nevertheless done regularly (as stipulated by law), are sea crew and ship passengers. Usually not stipulated by law, but nevertheless done occasionally, are control room operators (nuclear plants, electrical plants, chemical plants, etc.). Further down the line, we get to truck and bus drivers taking refresher courses in defensive driving; and train operators, whose occupation often produces major work accidents.

Indeed, our daily lives are peppered with industrial and home accidents, many of which may have been avoided if the parties involved were sufficiently experienced to behave and react correctly at the time of the potential accident.

In most instances, the circumstances surrounding a particular accident occur *very* infrequently. A passenger ship may never experience a major fire during its entire operating life. Most nuclear plant controllers may never experience the circumstances that occurred during the "3-Mile Island" disaster. How many operators have actually experienced a rotating device accident, in which items of the operator's clothing get entangled in the machinery? Thus, the probability of a dangerous situation occurring at all is usually low, and that this would develop into a real accident has an even lower probability.

We know enough about "Learning-Forgetting-Relearning" in order to evaluate the performance status of an operator, and to possibly evaluate the cost for raising the operator's performance to an acceptable level. Globerson et al. (1998) does just that in his experiments with fighter pilots.

What else we need, is to estimate the cost of a response error which results in some 'damage'. Armed with this data we would then be in a position to develop an Optimal Training Schedule, which provides an optimal balance between training costs and the potential damage that an accident might incur. Such a model will be developed based on research done by Altman and Dar-El (1998).

Along similar lines, in another research by Ginzberg (1998), the near optimal training conditions are determined experimentally. While we did not evaluate the probable cost of an accident, we did control the situation by not allowing the performance at the end of the forgetting period to get below a defined level.

In several respects, this work is similar to that done by Globerson et al. (1998). However, the situation Ginzberg investigated was a real task faced by the IDF (Israel Defense Forces) whose conclusions resulted in changes to the IDF's operator training schedules.

This chapter will be presented under the following titles:

7.1   Near Optimal Training Schedules via Experimentation: A Military Application.

7.2   Developing Optimized Training Schedules.

7.1 Near Optimal Training Schedules via Experimentation: A Military Application

A considerable number of studies on retraining and skill retention focus on the military work environment. This is due to the fact that many military tasks are of complex nature requiring high levels of skill on the part of their operators. Ordinarily, the operators do not perform the task often and therefore, refresher training is required in order to achieve skill retention. For example, the examination of the skill performance of helicopter pilots by Ruffner and Bickly (1984) indicates a critical decline in their flight skills after six months of no flying. Similar results were found by Wisher et al. (1992) by observing reserve soldiers before the Persian Gulf War.

Much of the relevant military literature is based on field experiments, but little is known in the civilian research community, since these internal reports are generally unavailable to the general public.

Ginzberg (1998), working on his M.Sc. thesis, planned his experiments on the following understanding/conclusion gleaned from the literature known at the time (about 1996):

(a)   Skills tend to decay with the passage of time

(b)   Increasing inactivity, due to the passage of time, requires a longer retraining period.

(c)   Using appropriate retraining material will eventually raise the skill performance to its original level.

(d)   The relearning time duration, to achieve original skill performance, is shorter than the original learning period (i.e., to get to the original level).

(e)   The first cycles of retraining have a major impact on the improvement (relearning) process.

(f)   Procedural skills demand longer training time than does psychomotor skills, but the latter skills are retained longer than procedural skills.

(g)   Automatic skills are retained longer than controlled skills.

7.1.1   The problem description.   The Israel Defense Force (IDF) maintains an operational unit for Electronic Warfare. The case study deals with this unit.

Using simulators for task learning is common, but using it for retraining is less common. There are only a few articles about simulator training method concerning skill retention. Gray (1979), O'Hara (1990), Lampton et al. (1991) and Semple et al. (1981) found that using a task-specific simulator is effective with the trainees.

Each training session for the reserve soldiers in this unit consists of three exercises called "links" which take about 30 minutes each. Both during and at the conclusion of each link, performances are examined by an experienced officer. Four operators work at specific roles as a team and their individual as well as team performances are evaluated while the "links" proceed. At the end of each link, each operator is interviewed separately and a performance grade awarded.

Currently, each reserve operator is required to appear once a month to undergo the relearning program.

Reserve operator ages range from 22 to 47 years, and all were at their peak operational performance during their 3 years of compulsory service in the IDF. The measure used for the operator's level of performance (OLP) is regarded as being both fair and very high, with a level of 90% and over, considered to be peak performance.

This army unit contains only 55 positions and there is very high motivation and competition for newcomers to enter the unit. Likewise, there is very high motivation for current members to remain in the unit.

7.1.2   The pilot study. A pilot study was conducted, aimed at examining and verifying the various assumptions made regarding the reserve operations and their performance levels.

The regular evaluation method already existed and this was used as an overall reference point for the new questionnaires that were being tested. The basic assumptions examined were:

- The complex tasks required of the operators to use both psychomotor and procedural behaviors, which the new questionnaires could identify and separate.
- The OLP, at the close of one training session, places the team at peak performance.   The drop in OLP, after the first "link" of the consequent session, represents the "forgetting".
- The OLP, at the close of each training session, reaches the same high level achieved at all previous sessions (i.e., reviewing the scores of the third "link"). The three "links" represent the "relearning".
- There is no relation between the operator's age (or seniority) with his professional performance.

The same five evaluators were used during the pilot study and during the entire set of experiments. They were given specialized instruction to ensure link consistency, having the same scenario or pattern, and to ensure that the questionnaires would be written with the same emphases.   To avoid any bias, the evaluations were taken after every link, on each operator, during the entire study period, which lasted six months.

The questionnaires went through several refinement procedures until these were readied for the experiment. In all, eight questionnaires were used so that 'psychomotor', 'procedural', 'automotive' and 'controlled' behavior could be measured. However, this detail is outside the intent of this report.

7.1.3   The study objectives.
   a)   To investigate the relationship between relearning and skill retention.
   b)   To consider the effect of varying the time duration between training.
   c)   To improve the efficiency of training the army unit.

7.1.4   The experimental plan.   Forty-eight reserve operators were involved in this study. These were grouped into three lots of 16; the first group was required to retrain after a one-month interval (as was the practice until the experiment began). The second lot was asked to wait two months before taking their retraining session, while the third lot was given a three-month break before undergoing the retraining session.

Each lot of 16 operators were sub-grouped into 4 teams of 4 persons each (each person in a team has his own specialized training) in order to have four samples for each condition tested. Figure 7.1 shows a layout of the experimental plan.

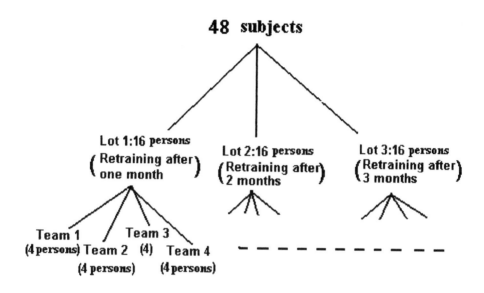

Figure 7.1: Layout of experimental plan

7.1.5   Partial experimental results. Most relevant to this study are the results shown in Figure 7.2, which show the average score values given after each "link" for the three interruption conditions. Link "0" is considered to be the starting point for retraining for each interruption condition, and is the 90% OLP score level obtained at the last training session, which is deemed as satisfactory (this may not be true for the second session for the 3-month interruption, since its end point at the end of the first session did *not* reach 90%).

Only with the interruption levels of 1 and 2 months, do we see a complete recovery to peak performance at the end of the third link (i.e., the end of the day's training). The 3-month interval case yields unsatisfactory results. Not only is peak performance not attained, the initial forgetting (at the close of link 1) shows a high loss of skills – too high to risk this performance level in a real emergency.

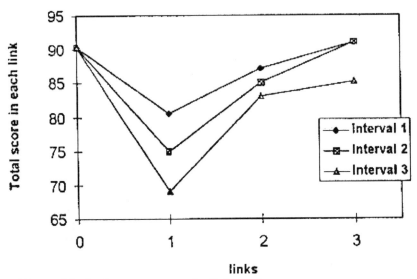

Figure 7.2: Performance versus the link number

The P value was tested between the total scores in link No. 3 of the three intervals (The T test and the SIGN test were both used). The results are summarized in Table 7.1.

According to Table 7.1, there is a significant difference between the OLP after 'link 3' between the training intervals of one or two months and the interval of three months. No significant difference was found between the OLP after link 3 for the one-month and the two-month intervals.

Table 7.1: The difference between the OLP-total score at the end of 'link 3' according to the training interval

Dimension/statistical	T test P value	SIGN test[*] P value
Interval 1 # Interval 2	0.432	0.8036
Interval 1 # Interval 3	0.000	0.000
Interval 2 # Interval 3	0.001	0.000

[*] In the SIGN test the P value was determined by the binomial distribution because the number of the participants was less than 25 (only 16 were used).

7.1.6   Improving the efficiency of the training schedule. Evaluating the results, shown in Figure 7.2, it seems a small price would be paid if the interruption level was increased from one month to two months (i.e., halving the reserve man-days). The price would be that performance score would drop from about 81% (currently experienced) to about 75% with a 2-month interval.

The military tends to be very conservative and, after consultations with the senior staff, it was agreed to make the change in steps, beginning with an interruption period of 1½ months, and to evaluate the results. Loss in performance would now be extrapolated and estimated at about 78%, with the certainty that a peak score of 90% would be achieved by the end of the training session (i.e., completion of 3 links).

Furthermore, during a real call-up, reserve operators usually have ample time to simulate at least 1 link to bring their performance to about 85% (assuming the call-up comes just before their regular retraining session).

The IDF has adopted these recommendations and, thus far, the training schedule operates as planned, with a total saving of one third of the current reserve operator man-days required (i.e., each reserve operator puts in 8 retraining days per year instead of the current 12 days a year).

7.2 Developing Optimal Training Schedules

The following is a summary of Altman and Dar-El's (1998) work contributed as independent researchers, within a research team, headed by Professor Dar-El, supported by the Ministry of Labor, for developing Industrial Optimal Training Schedules (see Dar-El and Zohar, 1998).

7.2.1 Introduction.
(One) Problem description: We consider typical industrial processes which are characterized by the learning phenomenon, wherein reductions in response time occur due to the repetition of some action performed a number of times. Training implies repetition and apart from the objective of achieving fluency in performance, it has an essential role to play in achieving improved responses at a time of crisis. It is generally impractical to train a crew of operators continuously over time. Break periods, other than those taken as meals, sleep, weekend, etc., often separate training sessions. It is well known that 'forgetting' takes place during break periods with performances after the break being of a poorer quality and/or slower execution time. However, at the

time of this research (1997), there were no theoretical models available of the learning-forgetting-relearning (L-F-R) phenomenon.

If operators need to be highly proficient in their work, then perhaps they will need to be continuously trained, but this could be exceedingly expensive. A far less expensive model is to give them plenty of training at the beginning, then leave them for an extended period before giving them some retraining to update their skills. The problem here is that an emergency response may be required during the 'extended break' but the actual responses may not be performed in good time. There is, therefore, a need to determine how often training periods are required in order to achieve sufficiently good responses without incurring high costs. This research is concerned with the development of a least cost effective training policy/training schedule, requiring the determination of the following factors: The initial training length, the retraining period's frequency, and the retraining period's length.

(Two) Concepts: We consider the Learning process, modeled by the power curve, as in equation (1.3a), which is rewritten below, *but for this section alone*, t_n becomes $Y_{(n)}$ and t_1 becomes $Y_{(1)}$. We do this because time t becomes a variable and will be used in developing the equations. Thus,

$$Y_{(n)} = Y_{(1)} \cdot n^{-b}$$

where,

$Y_{(1)}$ – is the performance time for the first repetition
$Y_{(n)}$ – is the performance time for the n-th repetition
b   - is the learning constant.

Call the Response Time $R(n)$, which is equal to Y_n when no breaks are present. We define the Failure Probability as the probability of failing to respond correctly during a crisis, given the response time at the crisis as $\omega(R)$. We utilize the concept of Acceptable Response Rate (ARR) to indicate the acceptable level for the acquisition of skills, so that in an emergency, an operator, whose performance is at least equal to the ARR, will respond correctly and in good time to the emergency. We define the Acceptable Response Rate (ARR) as follows: Given an acceptable failure probability α, we determine ARR by solving $\alpha = \omega(ARR)$.

(c) Solution Methodology: We will develop our solution methodology along the following lines:

- Build a model for learning with breaks.
- Empirically check the validity of this model.
- Build a cost function for a training schedule based on the above model. Include the trade-off between training costs and failure costs in the model.
- State the optimization problem.
- Develop a solution method for the optimization problem.

(d) Transducing the Power Curve Y_n as a Function of "n" to a Function of "t": $Y(n) = Y_{(1)} \cdot n^{-b}$, is not a continuous function in n , given we have interruptions in the learning process (see Figure 7.3).

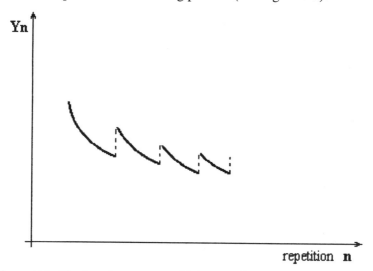

Figure 7.3: The learning process with interruptions

In order to get a continuous function, which is preferred for optimization problems, we need to consider 'time' as an independent variable for Y. How does $Y_{(n)}$ look as a function of time, t?

$$Y_{(n)} = Y_{(1)} \cdot n^{-b} \tag{7.1}$$

and the Production time for n cycles $\equiv T(n)$, where T(n) is the cumulative production time for n cycles. Hence,

$$T(n) \simeq \frac{Y(1)n^{1-b}}{1-b} \quad \text{for large n} \tag{7.2}$$

and $T(n)$ is equivalent to 't', the time axis. Thus, from (7.2) we get:

$$n(t) = \left[\frac{1-b}{Y(1)}\right]^{1/1-b} . t^{1/1-b.} \tag{7.3}$$

Substituting (7.3) in (7.1):

$$Y(n(t)) \equiv Y(t) = \underbrace{Y(1) \cdot \left[\frac{1-b}{Y(1)}\right]^{1/1-b}}_{K} \cdot t^{-b/1-b.} \tag{7.4}$$

Setting $\beta \equiv b/1-b$, we have:

$$Y(t) = Kt^{-\beta} \tag{7.5}$$

$Y(t)$ is an exponential function similar to $Y(n)$!

(Five)   *The Interrupted-Learning Curve (R(t)):*   Consider the building
   blocks of $R(t)$, the Reaction Time (or, the performance time at 't'):
   • The learning curve – is now a function of time $Y(t)$.
   • The "forgetting" curve, $g(t)$, which represents the increase in
     response time $R(t)$ due to a pause in the learning process.
   • The "relearning" curve, $h(t)$, which represents the difference from
     the response time $R(t)$ during retraining. This difference shrinks
     rapidly to zero as shown in Chapter 5.
     A typical "Interrupted-Learning-Relearning" curve is illustrated in
     Figure 7.4, where values for g and h can change for each retraining
     cycle.

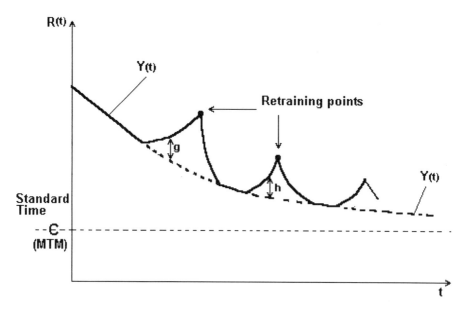

Figure 7.4: Illustrating the forgetting and relearning processes

In order to express $R(t)$ mathematically, we need to define some points in time.

Refer to Figure 7.5:

t^i – initial training length.

t_n^b – n-th retraining beginning time.

t_n^r – retraining length for n-th retraining cycle.

t_n^p – break length for n-th retraining cycle.

t_n^{r1}, t_n^{r2} – as shown in Figure 7.5.

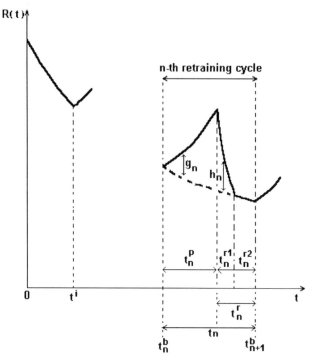

Figure 7.5: Details of the forgetting and relearning processes

Assuming that during all retraining cycles, we reach the original $Y(t)$, we can write the mathematical expression for $R(t)$:

$$R(t) = \begin{cases} Y(t) & ; \quad t \leq t^i \text{ or } t_n^b + t_n^p + t_n^{rl} < t \leq t_{n+1}^b \\ Y(t) + h_n(t - t_n^b) & ; \quad t_n^b < t < t_n^b + t_n^p \\ Y(t) + h_n(t - t_n^b - t_n^p) & ; \quad t_n^b + t_n^p < t \leq t_n^b + t_n^p + t_n^{rl} \end{cases} \quad (7.6)$$

7.2.2   The cost function.

(a) *General:* We will now build a cost function that expresses the cost per unit of time for a particular training schedule. A training schedule is completely determined by giving the values of the following variables: t^i, t_n^p, t_n^r for all n's in the planning horizon T. (Assuming we know $R(t)$ for any $t \leq T$). We distinguish between two phases:

(1) A learning phase – during which $R(t) \ll C$, the standard time.

(2) A steady-state phase – during which we have reached, or are close to reaching the MTM time for the task ($R(t) \sim C$).

The cost function will be developed for phase (1), the more general learning phase.

(b) *Cost Factors.* We identify three cost factors – the initial training cost, retraining cost and failure cost.

(a) *The Initial Training Cost (ITC):*

$$ITC = U + u \cdot t^i \tag{7.7}$$

U – constant initial training cost (e.g., the setup cost)
u – cost per unit of time for initial training.

(b) *Retraining Cost for the n-th Retraining Cycle (RTC_n):*

$$RTCn = V + v \cdot t_n^r \tag{7.8}$$

V – is a constant retraining cost (e.g., the setup cost).
v – is the cost per unit of time for retraining.

(Note: The model can be easily expanded for different retraining costs per retraining cycle by using V_n and v_n instead of V and v).

We assume V and v are independent of the initial response time at t_n^b.

(c) *Failure Cost (FC):* Let X be the random function for time to a crisis.

$fx \sim \text{Exp}(1/\lambda)$ – time to next crisis.

$\omega(R)$ is the probability of failure, once a crisis occurs, given a response time at the crisis R. The cost of failure over a planning horizon of T will be:

$$FC = \sum_{i=1}^{N_c} \omega(R(t_i^c)) \times D, \tag{7.9}$$

where

t_i^c – time of crisis i

N_c – number of crises that occur during T

D – the mean cost of failure at a crisis.

We can estimate this cost through simulation, or making some simplifications of the equation via linearity assumptions.

In general, $\omega(R)$ is a nonlinear function of R, as illustrated in Figure 7.6.

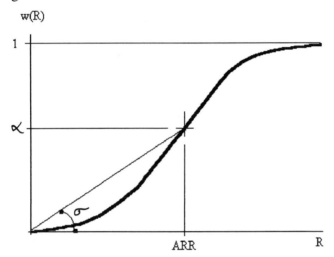

Figure 7.6: $\omega(R)$ as a nonlinear function of R

As R increases, $\omega(R)$ tends to 1 ($R \rightarrow \infty$ means no action was taken at the crisis point leading to a sure failure).

If we permit $\omega(R)$ to be no higher than α (and thus we set ARR), we can use a linear estimation for $\omega(R)$:

$$\omega(R) \sim \sigma \cdot R \qquad (7.10)$$

Thus, (7.9) becomes:

$$FC = \sum_{i=1}^{N_c} \sigma \cdot R(t_i^c) \cdot D. \qquad (7.11)$$

The expected mean, E(FC) is then given by:

$$E(FC) = (\lambda \cdot T) \cdot \sigma \overline{R}(t) \times D = \lambda \cdot \sigma \cdot D \cdot T \cdot \overline{R}(t) \equiv z \cdot \overline{TR}(t).$$

$$(7.12)$$

This Failure Cost can be separated into two components: a base failure cost (BFC), which is the cost that is derived from response time without interruptions $(Y(t))$ and an extra failure cost (EFC), which is derived from increases over $Y(t)$ caused by interruptions.

Given a planning horizon T, BFC is not affected by the training schedule. Thus, it can be omitted (BFC is equal to $z \cdot T \cdot \overline{Y}(t)$ or $z \cdot \overline{Y}(t)$ per unit of time).

EFC increases as we increase the length of learning interruptions, since this increases $R(t)$ which, in time, increases failure probability and thus the mean failure cost. So, the trade-off emerges: Increase interruptions to save training money, at the cost of increased failure risk.

The equation for the extra failure cost:

$$EFC_n = z \cdot T \cdot (\overline{g}_n \cdot \frac{t_n^p}{T} + \overline{h}_n \cdot \frac{t_n^r}{T}) \qquad (7.13)$$

(d)   *Linear Approximation for $g_n(t)$ and $h_n(t)$:* In order to further simplify our model, we can make some assumptions for g and h. We assume g_n is a function of: the interruption length, the initial response time at the break $(Y(t_n^b)$ and MTM time of the task, C. The validity of these assumptions will be checked empirically through experiments that will be described later. Thus,

$$g_n(t) = \delta_n \cdot t \qquad (7.14)$$

where,
t – is the length of interruption
δ_n – is a proportionality constant, forgetting slope, where

$$\delta_n = \delta \cdot \frac{Y(t_n^b)}{C}, \qquad (7.15)$$

where,

δ is the minimal forgetting slope. In this model, δ_n, the forgetting slope, decreases as we learn more and approach C.

$$g_n(t) = \delta \cdot \frac{Y(t_n^b)}{C} \cdot t \tag{7.16}$$

ERRORS in eq'n (7.17)

$$\bar{g}_n = \frac{1}{2} g_n \cdot t_n^p \frac{\delta t_n^p}{2C} \cdot Y(t_n^b) \cdot t_n^p = \frac{\delta t_n^p}{2C} \cdot K \left(t^i + \underbrace{\sum_{j=1}^{n-1}(t_j^p - t_j^r)}_{=t_n^b} \right)^{-\beta} \tag{7.17}$$

Similar to g_n, we can approximate h_n by:
ERRORS in (7.18)

$$h_n(t) = \begin{cases} g_n(t_n^p) - \gamma_n \cdot t \, ; 0 \leq t < t_n^{rl} \\ 0 \qquad\qquad\quad ; t_u^{rl} \leq t \end{cases} \tag{7.18}$$

where γ_n is given by:

$$\gamma_n = \gamma \times \frac{g_n(t_n^p)}{C} \tag{7.19}$$

γ - is the minimal relearning slope.

$h_n(t)$ begins at a value of $g_n(t_n^p)$ and becomes zero at a rate of γ_n. It remains zero afterwards. In order to calculate $\bar{h}_n$, we need to find t_n^{rl}:

$$h_n = 0 \Rightarrow t_n^{rl} = \frac{g_n(t_n^p)}{\gamma_n} = \frac{g_n(t_n^p)}{\gamma \times \frac{g_n(t_n^p)}{C}} = \frac{C}{\gamma} \ . \tag{7.20}$$

Thus,

$$\bar{h}_n = \frac{1}{2} g_n(t_n^p) \cdot \frac{t_n^{rl}}{t_n^r} = \bar{g}_n \cdot \frac{t_n^{rl}}{t_n^r}$$

$$\tag{7.21}$$

$$= \frac{\delta t_n^p}{2\gamma t_n^r} \cdot K \left(t^i + \sum_{j=1}^{n-1} (t_j^p + t_j^r) \right)^{-\beta}$$

(e) *The Total Cost Functions:*
(One) **The General Cost**

$$TC_1 = \frac{1}{T} \cdot \left\{ ITC + \sum_{i=1}^{N} (RTC_i + TFC_i) \right\}$$

where N, is the number of retraining periods during T.

$$TC_1 = \frac{1}{T} \left\{ U + ut^i + \sum_{i=1}^{N} (V + v \cdot t_i^r) + \sum_{i=1}^{N_c} \omega(R(t_i^c)) \times D \right\}$$

$$\tag{7.22}$$

(Two) **The Cost Function with Linear Approximation for $\omega(R)$:**

$$TC_2 = \frac{1}{T} \left\{ U + ut^i + \sum_{i=1}^{N} (V + v \cdot t_i^r) + z \cdot T \cdot \bar{R}(t) \right\} \tag{7.23}$$

(c) The Cost Function with Linear Approximation for g_n and h_n:

$$TC_3 = \frac{1}{T}\left\{ U + ut^i + \sum_{i=1}^{N}(V + v \cdot t_i^r) + z \cdot T \cdot Y(t) \right.$$

(7.24)

$$\left. + z \cdot \sum_{i=1}^{N}\left[\frac{\delta}{2} Y(t_i^b) \cdot t_i^p \left(\frac{t_i^p}{C} + \frac{C}{\gamma} \right) \right] \right\}$$

7.2.3 Defining the optimization problem. We can define three different optimization problems, each one corresponding to one of the cost functions described above. The constraints would be:

$$T = t^i + \sum_{i=1}^{N}(t_i^p + t_i^r)$$

(7.25)

and the decision variables are: t^i, t_i^p, t_i^r and N, all greater than zero.

(One)   Minimize equation (7.22)
s.t.: equation (7.25) is satisfied.
This problem would be solvable only by simulation.

(Two)   Minimize equation (7.23)
s.t.: equation (7.25) is satisfied.
This problem can be solved using the following method, given that g_n

and h_n are differentiable:

- Find equations for g_n, h_n.
- Calculate $\overline{g}_n, \overline{h}_n$.
- Differentiate equation (23) with respect to t^i, t_n^p, t_n^r.
- Solve the resulting equation system, with N as a parameter.
- Determine optimal N by further differentiation with respect to N.

Another possible solution method would be to develop and run a GAMS model.

(Three) Minimize equation (7.24)
s.t. equation (7.25) is satisfied,

$$t_i^b = t^i + \sum_{j=1}^{t-1} (t_j^p + t_r^j);$$

$$t^r \geq \frac{\delta_n}{\gamma_n} t_n^p (= t_n^{rl})$$

This problem is solvable by the same methods described above.

7.2.4 Analytical solution of a simplified model. In order to achieve some insights of the mathematical behaviour of optimal training schedules, we will now further simplify the above models and solve this simplification analytically.

We will assume g_n and h_n to be linear and independent of response time at the beginning of the break period. In fact, we use equations (7.14) and (7.18), for δ_n and γ_n constants which depend only on n. We then have:

$$\overline{g}_n = \frac{1}{2} \delta_n t_n^p \tag{7.26}$$

and

$$\overline{h}_n = \overline{g}_n \cdot \frac{t_n^{rl}}{t_n^r} \tag{7.27}$$

since h_n is zero for $t_n^{rl} \leq t < t_n^r$.

Using $t_n^{rl} = \dfrac{g_n(t_n^p)}{\gamma_n} = \dfrac{\delta_n}{\gamma_n} t_n^p$, we get:

$$\overline{h}_n = \overline{g}_n \cdot \frac{\delta_n}{\gamma_n} \frac{t_n^p}{t_n^r}. \tag{7.28}$$

Our cost function will be (substituting equations (7.26) and (7.28) in (7.13)):

$$TC_4 = \frac{1}{T}\left\{ U + ut^i + \sum_{i=1}^{N}(V + vt_i^r) + zT\overline{Y}(t)) \right.$$

$$\left. + z \cdot \sum_{i=1}^{N}\left(\frac{1}{2}\delta_i t_i^p \cdot t_i^p + \frac{1}{2}\delta_i t_i^p \cdot \frac{\delta_i}{\gamma_i} t_i^p \right) \right\}$$ (7.29)

Using (7.25) for t^i and $\overline{Y}(t) = \dfrac{KT^{-\beta}}{1-\beta}$ we get:

$$TC_4 = \frac{U}{T} + u - \frac{u}{T}\sum_{i=1}^{N}(t_i^p + t_i^r) + \frac{NV}{T}$$

$$+ \frac{v}{T}\sum_{i=1}^{N}t_i^r + \frac{zK}{1-\beta}T^{-\beta}$$

$$+ \frac{z}{2T}\cdot\sum_{i=1}^{N}\left[\delta_k (t_k^p)^2\left(1 + \frac{\delta_k}{\gamma_k} \right) \right]$$ (7.30)

By setting $\dfrac{\partial TC_4}{\partial t_i^p} = 0$ we get:

$$t_i^{p^*} = \frac{u}{z\delta_i\left(1 + \dfrac{\delta_i}{\gamma_i} \right)} \quad ; \qquad i = 1,\dots,N$$ (7.31)

This determines the optimal break times for the training schedule. We need to now find $t_i^{r^*}$:

$$\frac{\partial TC_4}{\partial t_i^r} = \frac{-u}{T} + \frac{v}{T} = 0$$ (7.32)

Since t_i^r is linear in TC_4, its optimal values will be at extremes. If $u > v$, meaning initial training is more costly than retraining (per unit of time), we prefer t_i^r to be as large as possible. Conversely, if $u < v$, we need the minimum value of t_i^r. Thus, if

$$u > v, t^{i^*} = 0,$$

$$\sum_{i=1}^{N} t_i^{r^*} = T - \sum_{i=1}^{N} t_i^{p^*},$$

or

$$u > v : t^{i^*} = 0,$$

$$\sum_{i=1}^{N} t_i^{r^*} = T - \frac{u}{z} \sum_{i=1}^{N} \frac{1}{\delta_i \left(1 + \dfrac{\delta_i}{\gamma_i}\right)}. \tag{7.33}$$

Distribution of the total retraining time, $\sum_{i=1}^{N} t_i^{r^*}$ among individual t_i^r's does not affect the total cost. This is true because of our simplifying assumption of δ_n, γ_n being constant. If there is another constraint for t^i (say $Y(t^i) \leq ARR$), we should set t^{i^*} at its minimum possible value and solve again for $t_i^{r^*}$:

$$u > v : t^{i^*} = t_{min}^i,$$

$$\sum_{i=1}^{N} t_i^{r^*} = T - t_{min}^i - \frac{u}{z} \sum_{i=1}^{N} \frac{1}{\delta_i \left(1 + \dfrac{\delta_i}{\gamma_i}\right)} \tag{7.34}$$

When $u < v$, $t_i^{r^*}$ must be as short as possible. Since we return to the power curve after each pause, we have:

$$t_i^r \geq \frac{\delta_i}{\gamma_i} t_i^p \qquad (t_i^r \geq t_i^{r1}) \tag{7.35}$$

So, we get for the optimal solution:

$$u < v : t_i^{r^*} = \frac{\delta_i}{\gamma_i} t_i^{p^*} = \frac{\delta_i}{\gamma_i} \cdot \frac{u}{z\delta_i\left(1+\dfrac{\delta_i}{\gamma_i}\right)} = \frac{u}{z\gamma_i\left(1+\dfrac{\delta_i}{\gamma_i}\right)} \qquad (7.36)$$

and

$$u < v: \ t^{i^*} = T - \sum_{i=1}^{N}\left(t_i^{p^*} + t_i^{r^*}\right)$$

$$= T - \frac{u}{z}\sum_{i=1}^{N}\left[\frac{1}{\left(1+\dfrac{\delta_i}{\gamma_i}\right)}\left(\frac{1}{\delta_i}+\frac{1}{\gamma_i}\right)\right]$$

$$= T - \frac{u}{z}\sum_{i=1}^{N}\frac{1}{\delta_i} \qquad (7.37)$$

For $u = v$, we can choose t^i and t_i^r as we prefer, as long as we have (7.25).

Optimal N is dependent on the functions of i, δ_i and γ_i. By substitution of optimal values in TC_4, and the calculation of the series for δ_i, γ_i, we eventually will have an expression for total cost as a function of N, thus permitting calculation of N^*.

7.2.5 Empirical determination of g and h. Experiments were conducted in order to empirically determine values for g and h. Four sets of simple assemblies were used and subjects were asked to assemble the work under varying interruption periods. Experimental details are fully reported in Dar-El and Zohar (1998).

7.2.6 An illustrative example.

Determining the optimal training schedule for forest fire fighters.
Suppose the cost of not reacting swiftly when a forest fire occurs gets to be very high. Let us assume that if the response time is 30 minutes, we have damages of 50 M\$, and the probability to incur these damages is linear with response time (linear approximation of $\omega(R)$ – see eq. (7.10)). Let us further assume the following figures:

Number of forest fires per year, $\lambda = 2$.
Cost of initial training of firefighters,   n = 70 K$/week.

Initial training time, t^i = 50 days.
Cost of retraining,   v = 50 K$/week.
Forgetting slope, δ_i = 0.1 min/week.

Remembering slope, γ_i = 0.2 min/week.

We would like to determine the optimal retraining schedule to minimize costs over a planning horizon T of five years. That is, we need to determine t_i^{p*}, t_i^{r*} and N^* (in this example, t^{i*} is set to a t_{min}^i of 50 days due to the initial training constraint and the fact that u > v − eq. (7.33)).

First, we calculate σ using eq. (7.10)

$$\omega(R) \cong \sigma R => \sigma. = \frac{1}{30} \text{ per min. } = 336 \text{ per week}$$

Then, using equation (7.12), we get:

$$z \equiv \lambda \cdot \sigma \cdot D = \frac{2}{52} \cdot 336 \cdot 50 \times 10^3 \cong 646 \times 10^3 \left[\frac{\text{K\$}}{\text{wk}^2} \right]$$

We can now use equation (7.31) to determine the optimal break periods:

$$t_i^{p*} = \frac{u}{z\delta_i \left(1 + \dfrac{\delta_i}{\gamma_i} \right)} = \frac{70}{646 \cdot 10^3 \times 0.1 \times \left(1 + \dfrac{0.1}{0.2} \right)} =$$
$$= 7.22 \text{ wks} \cong 50 \text{ days / cycle}$$

(For simplicity purposes, we have set all δ_i's and γ_i's equal for each i, with no loss of generalization.) Using equation (7.33), we get:

$$\sum_1^N t_i^{r*} = T - t_{min}^i - \frac{u}{z} \sum_1^N \frac{1}{\delta_i \left(1 + \dfrac{\delta_i}{\gamma_i} \right)} =$$
$$= 5 \times 52 - 7 - 7.22 \, N$$

If there are external constraints for N, for example, we have five retraining cycles per year. Then, we can use this to calculate t_i^{r*} directly:

$$\sum_{1}^{5} t_i^{r*} = 253 - 7.22 \times 5 \cong 500 \text{ days/planning horizon} \qquad (7.38)$$

This would give us the following training schedule:

- Initial training of 50 days.
- Breaks of 50 days between training cycles (t_i^{p*}).
- Retrain times of 20 days each cycle.

If we can choose N freely, then we should substitute eq. (7.38) into eq. (7.30) and divide by N to get the optimal number of cycles – N*. Then we would recalculate $t_i^{r^*}$ using equation (7.38).

7.2.7 Conclusion. This is the first attempt at developing a cost model for Optimal Training Schedules. Much more work still needs to be done in understanding the behavior of the 'g' and 'h' functions under general operating conditions – with their values verified by experimental methods. On the other hand, what value should be placed on 'life'? We have ignored all moral issues – but this will need to be considered.

To be pragmatic means we should assume reasonable values for these problematic items and apply the theory to achieve a first cut estimation of the optimal training schedule. Develop a feasible plan and operate the training system, but keep good records over time. Use whatever methods are available to evaluate the 'g' and 'h' factors and recalculate the optimal training schedule. If needed, make adjustments to your current schedule and continue with the 'next round' and so forth.

Finally – Please Publish Your Findings!

8 LEARNING IN THE PLANT: THE BEGINNINGS OF ORGANIZATIONAL LEARNING

In this chapter, we cover the application of learning curve theory to the firm as a whole. We call this type of learning *Progress Functions* to distinguish it from *Individual Learning* (see Chapter 1). Application areas differ considerably; some deal with 'aggregate planning'; some with 'capacity planning' project scheduling, while others discuss the influence of quality (TQM), as well as JIT (Just-in-Time). The latter two items, especially, begin to encroach into 'organizational learning' and therefore will appear in the latter part of this chapter, since Chapter 9 deals with the learning organization.

Many of the topics are concerned specifically about the subject on which they are written, and 'learning theory' is only an incidental part. In those cases, it is not intended to go into great depth in the respective subject matter, but rather to show how learning is taken into account.

For want of a better location in this book, some references to 'econometric modeling' with the learning curve is included in this chapter. Again, the topic is merely introduced and left at that. So too is the vast literature on machine learning which appear in Computer Science and IEEE journals, but which also make liberal use of LC theory and jargon.

Total product learning is certainly a good introduction to 'learning in the plant', except that not much research has been done on this subject in the last 25 years. Remember, Wright's original power curve (Section 3.1), the Cumulative Average Power Curve (Section 3.2) and the Standard B Learning Curve (Section 3.3) were all developed for whole products (the airplane was *the* product). Obviously, both organizational and individual learning factors were simply lumped together and assumed to somehow operate in practice. Included here are the works by Crawford (1944), Titleman (1957), Nadler and Smith (1963), Glover (1966a, 1966b, 1967), Sahal (1979), Adler (1990), Hayes and Clark (1985), Montgomery and Day (1985), Alden (1974), Liao (1979), and Li and Rajagopalan (1997).

The topics discussed in this chapter are as follows:

8.1 Learning at the firm level: The progress function
8.2 Learning with new processes and products

8.1 Learning at the Firm Level: The Progress Function

Several papers have appeared which deal with the influence of learning on new processes, products and on characteristics of the firm itself. The early work by Wright (1936) was, in fact, applied to large products (aircraft) and the learning slope of 80% was based on these products. It seems that the aircraft industry had an exclusive hold on the learning curve until after WWII (see Asher, 1956; Carr, 1946; Chasson, 1945; Hirsch, 1952; Middleton, 1945; Garg and Milliman, 1961; Hartley, 1965; and Rapping, 1965). But, applications in other industries were not long in coming. These included shipping, electronics, machine tools, EDP system components, musical instruments, papermaking, steel, apparel, rayon, and automobiles (see Alchain, 1950; Baloff, 1966a, 1966b, 1970, 1971; Billon, 1966; Conway and Schultz, 1959; Hirsch, 1956; Searle and Goody, 1945; Hollander, 1965; Hufbauer, 1966; Joskow and Rozanski, 1979; Harvey, 1981; Sharp and Price, 1990; Hackett, 1983; and Zimmerman, 1982).

Conway and Schultz (1959) make an excellent presentation on the problem faced by researchers working on products in an industrial setting. They begin by identifying many of the major problems of data accuracy and data consistency that most researchers are likely to face. So enormous are these problems, that one wonders about the reliability of *any* of the industrial data reported in the professional literature.   Such problems include: the data availability format; design changes; paced and unpaced conditions, lot sizing (forgetting was 'unheard of' at the time); aggregation of information – problems with separating out aggregate and cumulative data; time delays in obtaining costs; and so on. Baloff (1971) also refers to the credibility of available data from the aerospace and other industries "...in which defense contracting is pervasive..." (see also Young, 1966; Zieke, 1962; and Globerson and Levin, 1995).

Conway and Schultz (1959) and Baloff (1966, 1971) provide some of the best examples we have *to date* on learning data drawn on log-log scales with fairly good explanations as to why the shapes turn out the way they do. These are essentially "Field Research" and little could be learned of the underlying learning concepts – other than they used the power model for all their analyses.

Table 8.1 is a summary of Baloff's (1971) experimental data taken in three industries: musical instruments, apparel and automotive. Only data relevant to this work is included.

Table 8.1: Regression Results (from Baloff, 1971**)

Industry	Product	t_1	b	(ϕ)	R
Musical	M1	240	0.191	(87.5)	0.97
Instruments	M2	199	0.188	(87.8)	0.98
	M3	218	0.175	(88.7)	0.90
	M4	59	0.173	(88.7)	0.94
	M5	93	0.141	(90.5)	0.97
	M6	3200	0.868*	(54.8)	0.99
Apparel	A-1	161	0.353	(78.3)	0.90
	A-2	123	0.333	(79.4)	0.96
	A-3	377	0.409	(75.3)	0.97
Automotive	AG-1	514	0.338	(79.9)	0.96
	AG-2	180	0.258	(83.6)	0.96
	AG-3	65	0.179	(89.2)	0.96
	SM-4	31	0.145	(90.4)	0.86

**Use of copyright material by Operational Research Quarterly is gratefully acknowledged.
*This was a large instrument that was 'constructed'. Learning was extremely rapid with a sudden steady-state occurring around n=30. It would have been useful to know the incentive pay earning for this group compared to those working on M1 to M5.

Regression coefficient values were very high, and with the exception of M6, the 'b' values more or less clustered about a central value for each industry. This led Baloff (1971) to speculate that there could be "typical" values of 'b' for particular industries. But, there are always exceptions to the rule, which could make a complete mockery of predictions. Baloff even tried to correlate t_1 and 'b' on a general level, then on an industry level, but no relationship was forthcoming.

The analogy to individual learning is very strong. What Baloff needed to do was to divide the t_1 values for each product by the appropriate MTM estimate and plot this with respect to the appropriate ϕ value. This writer is convinced that a fairly clear linear relationship would develop, and at once, we would have a reliable, stable and universal relationship for linking t_1 with b, or (ϕ).

This we will leave to future researchers to validate.

However, the problem of the power curve is not that simple to explain away. Even for individual learning, where support for the power curve is fairly solid, the question arises as to why or how learning is explained by the power curve. We have already concluded that the improvement process comes via many guises, e.g., skill acquisition, method improvement, simple design changes, etc. Is each element supposed to improve learning via the power curve? Obviously not! If it did, then we could not create a combined power curve for the whole job, since

$$t_n = \sum_{i=1}^{k} t_{1_i} \cdot (n)^{-b_i} \neq \hat{t}_1 \cdot n^{-\hat{b}}$$

(8.1)

where,

$$\hat{t}_1 = \sum_{i=1}^{k} t_{1_i}$$

and

$$\hat{b} = b_1 + b_2 + \cdots + b_k$$

there being k elements that contribute to the learning.

Perhaps the only explanation is that the combined effects of all these learning elements lead us to the single expression of the power curve.

The problem reappears when dealing with the assumption of a power curve for an entire product (as Wright and others did). Wright and others used the simple power curve to represent learning of an entire product (see also Kiechel, 1981; Day and Montgomery, 1983). What of the major parts and subassemblies of the product – do these not follow the power model? Impossible! If they do, then we cannot describe learning of the whole product as a power curve.

There seems to be little way out of this dilemma. However, Kantor and Zangwill (1991) develop a Learning Rate Budget (LRB) model to get around this problem. The LRB describes the learning curves for each major component of the product and in this way, facilitates planning, budgeting, control and management. The authors develop a somewhat complex theory which they claim is applicable to real manufacturing data. Their claim still needs verification, but without a doubt, shows promise as a method for analyzing large products.

Dutton and Thomas (1984) and Dutton et al. (1984) present excellent papers on managerial opportunities afforded via Progress Functions. They see it as an especially powerful tool for managers in competitive environments (see also Abell and Hammond, 1979; Fudenberg and Tirole, 1983; Cunningham, 1980; Hofer and Schendel, 1978; Gold, 1981; Bass, 1980; Conley, 1976; Doland and Jeuland, 1981; Dada and Srikanth, 1990; Enis, 1980; Clarke et al., 1982; Spence, 1981; and Howell, 1980). Dutton and Thomas' (1984a) progress functions are based on the power curve, where cumulative volume (a proxy for experience) is the only input variable. Increasing the rate of output leads to increasing unit cost reductions (Abell and Hammond, 1979; McIntyre, 1977 and the Boston Consulting Group, 1970). However, others have questioned the validity of the market share effect (see Hall, 1980; Woo and Cooper, 1982; and others), arguing "... that high performance via low cost produces advantages does not necessarily correlate with high market share". Nevertheless, sustained production often provides producers with opportunities to effect cost efficiencies that have little to do with cumulative volume.

Table 8.2 summarizes learning slopes of the Progress Functions for several studies.

Some slopes appear to be very sharp (see the first few items in Table 8.2). It is most unlikely that such values for the learning slope would occur with individual learning (compare with values found in Tables 4.1 – 4.3). This is because design changes, quality and productivity improvements are diligently practiced by many organizations, ensuring a continual improvement in a company's operations.

Additional data on progress function learning slopes are shown in Figure 8.1, based on Dutton and Thomas' (1984a) study of learning slopes (or, progress ratios) of over 100 studies whose learning slopes (neglecting the two 'pathological' extremes) range from about 61% to 96% which is quite reasonable for whole products.

Table 8.2: Progress Function slope values.

Application	Learning Slope
Production Man-Hours/ton of steel (Hirschman, 1973)	60
Ass'y electro-mech. products (Conway & Schultz, 1959)	68 - 75
Semiconductors production (Gruber, 1992)	68 - 75
Machine tool assembly (7 models) (Nadler & Smith, 1963)	69 - 73
Avge. revenue/silicon transistor for industry (Conley, 1976)	70
Price/pound polyvinyl chloride (const $) (Conley, 1976)	70
Price/unit – integrated circuits (Conley, 1976; Noyce, 1972)	70 - 72
Overhead cost/airplane (Wright, 1936)	70
$/ton – coal (const $) (Fisher, 1974)	75
$/kw-hr (const. $) (Fisher, 1974)	75
MH/barrel refined petroleum in USA (Hirschman, 1973)	75
Maintenance MH shutdown/GE Plant (Hirsch, 1956)	75
Prod'n hrs/Liberty ship, tankers, etc WW2 (Hirsch, 1956)	78 - 84
$/barrel gasoline USA (Hirschman, 1973)	80
MH/airframe (WW2) (Hirsch, 1956)	80
Labor Cost/airplane (Wright, 1936)	80
Total labor/unit (16 machine tools) (Hirsch, 1956)	82
$/Model T Ford (1910 – 1926) (const $) (Hirschman, 1973)	86

The distribution of slopes in Figure 8.1 also shows the danger inherent in fixing "ϕ" for all learning, as Wright does at 80% (see also Globerson and Crossman, 1976; and DeJong, 1967). Instead, Dutton and Thomas recommend that ϕ should be a variable, whose value "… is controllable via creative managerial efforts", especially ones involving investments, in particular niche opportunity areas. Thus, the slope ϕ is more an outcome of managerial policy decisions. The slope may depend partly on production decisions, partly on marketing decisions and partly on joint decisions. The question, then, immediately shifts to the issue of *how* the rate of improvement can be managed. For example, capacity planning, sales forecasting, production planning, quality assurance and new product introductions are typical areas in which interfaces are needed on a routine basis between marketing and production managers (see also Hayes and Wheelwright, 1979; McCann and Galbraith, 1981; Skinner, 1969; and others).   It is doubtful that marketing and manufacturing, working independently, could induce high rates of cost improvements.

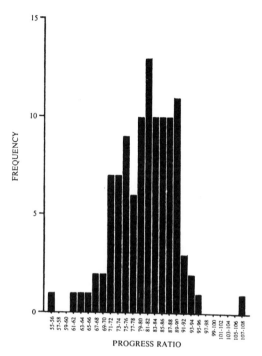

Figure 8.1: Distribution of progress ratios (N = 108)
(source: Dutton and Thomas, 1984)

According to Dutton and Thomas (1984a, 1984b), there is strong evidence to show that cumulative investments were superior in supporting Progress Functions (Arrow, 1962; Sheshinski, 1967, Hollander, 1965; and others), than improvements in direct labor.   In fact, greater inputs to the Progress Function could come from indirect labor, via improved scheduling, inventory control, wage incentives and via changes in product design.

Until now, we dichotomized individual learning as "autonomous" or "induced".   This 2-way classification is no longer adequate for Progress Functions.   Instead, Table 8.3 is reproduced below to show Dutton and Thomas' 4-way classification of the Learning Types.

Table 8.3: Some Examples of Four Learning Types via Which Firms May
Capture Progress Effects (after Dutton and Thomas, 1984a)

Autonomous Learning	*Induced Learning*
Exogenous origins	
1. General growth in scientific and technical knowledge that flows freely into the firm (Nelson and Langlois, 1983);	1. Learning of capital goods' suppliers induced by the users' experience with the equipment (Joskow & Rozanski, 1979; Von Hippel, 1976);
2. Continuously improving productivity garnered when a firm periodically replaces its equipment (Arrow, 1962).	2. Investment in improved capital goods in order to hasten the rate of progress (Hollander, 1965; Searle and Goody, 1945);
	3. Copying and adapting the technological innovation of a successful competitor (Mansfield, 1961; Tinnin, 1983).
Endogenous origins	
1. Direct-labor learning due to the "practice-makes-perfect" principle or wage-incentive plans (Conway and Schultz, 1959; Lundberg, 1961);	1. Improved tooling (Chassen, 1945; Conway and Schultz, 1959; Wright, 1936);
2. Routine production planning (Baloff 1966b, 1970; Conway and Schultz, 1959; Nadler and Smith, 1963).	2. Manufacturing process changes (Crawford and Strauss, 1947; Middleton, 1945);
	3. Model or product design changes to effect efficiencies in production (Billon, 1966; Conway and Schultz, 1959; Wright, 1936).

Reviewing Table 8.3, we see that Direct-labor learning (Endogenous #1)
is the *only* item that is obviously attributed to labor – the other nine points
have nothing to do with individual learning – but their impact is seen as
"Learning"!

Dutton and Thomas (1984a) conclude that "Without an understanding of
how different casual factors interact and influence the firms' cost dynamics,
prescriptions for using the progress principle have limited value.   The

literature lacks longitudinal studies that control for different factors, thus isolating relative effects. The state-of-the-art does yield a framework for managerial use in exploring different bases of progress and is considering impacts of organization on progress effects".

In another innovative study, Adler and Clark (1991) report on a partial analysis of Dutton and Thomas' (1984a) 4-way classification of learning types, by focusing on the "autonomous/induced" dimension, using data collected over the first three years of a hi-tech company's new generation product.

Two learning models are tested. The first is the regular power learning curve model (in its natural logarithm format), and in which productivity is equated to cumulative experience, and the second, a "Learning Process Model".

The Learning Process Model works on the assumption that a significant part of the productivity in the learning curve model is due to the influence of identifiable managerial actions, such as "Engineering Changes (EC)" and training. They separate first-order learning from second-order learning – after Dutton and Thomas (1984a) and Fiol and Lyles (1985).

Engineering Changes (EC), design errors, market experience, ease of manufacture (producibility) and cost reduction outcomes are all included. Training includes direct personnel who were obliged to undergo instruction and 'training' each time an EC was initiated. In addition, training included the regular instruction given for on-the-job activities.

Figure 8.2 is the Learning Process Model used by Adler and Clark (1991), which has two distinct paths – one corresponding to first-order autonomous learning, the other involving induced learning of which EC and training are the main second-order learning ingredients.

The analysis of their results produced some very insightful observations. For example, they found that the learning effect can be just as strong in very capital-intensive operations as in labor, or material-intensive operations. This finding is contrary to a common misconception that the opposite is true (they cite a DoD: DCAA (1970) report of some 182 learning curves for which the proportion of variance in learning slopes explained by the single variable "% of assembly work in total" was a meager 2.9%).

A second conclusion was that the learning processes are quite dissimilar across departments in the same company. This is not a very surprising conclusion. The two departments that were investigated were very different. One was highly capitalized and had a large complement of professional engineers, while the second was essentially a manually-oriented production process. The contributions of EC and training to productivity appeared to 'reverse' for the two departments and complex explanations were developed to explain this phenomenon.

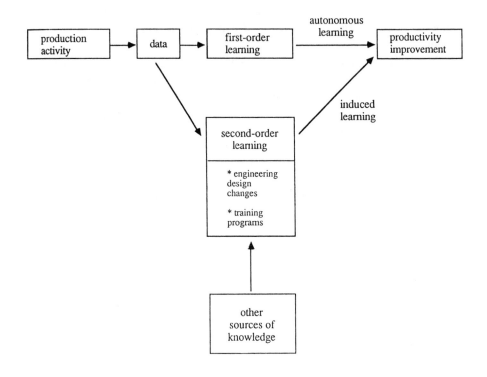

Figure 8.2: The learning process model

**Reprinted by permission, The Institute for Management Science (currently INFORMS), 901 Elkridge landing Road, Suite 400, Linthicum, MD 21090-2909, USA.

These conclusions also meant that the Learning Model shown in Figure 8.2 was inadequate, and a much more complicated model was needed.

This appears to be a good place to finish the discussion on Learning Models for the firm/products.   However, the paper on Learning on Production Processes by Levy (1965), although published 35 years ago, is still considered a "classic". Regrettably, its concepts' approach was simply not sufficiently developed by later researchers. Levy's work is discussed in the next section.

8.2 Learning with New Processes and Products

Research on new processes and products was tackled by Levy (1965), Howell (1980) and by Towill and his many collaborators. These approaches will now be considered. The topics will, therefore, be discussed under the names of the major authors.

8.2.1   Ferdinand K. Levy: Adaptation in the production process.
Levy (1965) begins with the traditional power curve (Equation 3.1a)

$$t_n = t_1 \cdot n^{-b} \tag{3.1a}$$

This expression is converted to the 'rate of production' R(n), where

$$R(n) = \frac{1}{t_n} = \frac{n^b}{t_1} \tag{8.2}$$

Levy was aware that R(n) asymptotes to infinity and consequently, assumes that the process has a maximum rate of output P (This was very perceptive, predating the work of DeJong and Towill who made similar assumptions.). The firm's cumulative experience (or, stock of knowledge) for a particular job would be based on the total output of the job up to that point in time.

Until the maximum output rate, P, is achieved,

$$R(n) < P \tag{8.3}$$

Levy assumed that the rate of increase in the production rate is proportional to the amount that the process can improve (reasonable assumption). Thus,

$$\frac{dR(n)}{dn} = \mu\{P - R(n)\} \tag{8.4}$$

The value of 'μ' depends on the particular process and is referred to as the process' "Rate of Adaptation". Solving equation (8.4) for R(n), we obtain

$$R(n) = P\left\{1 - e^{-(a+\mu \cdot n)}\right\}, \tag{8.5}$$

where, a is a constant of integration, and indicates the initial efficiency of the process, when n = 0, i.e.,

$$R(0) = P\left\{1 - e^{-a}\right\} \tag{8.6}$$

Levy then postulates that μ, the rate of adaptation, is influenced by several variables that are exogenous of n, denoting these as $y_1, y_2, \ldots, y_m$, so that

$$\mu = f(y_1, y_2, \ldots, y_m) \tag{8.7}$$

Assuming that μ is at least twice differentiable, equation (8.7) can be expanded in a Taylor Series, and dropping all terms higher than the first order, we get

$$\mu = B_b + \sum_n B_i n_i \tag{8.8}$$

where B_i's are constants.

By substituting μ from equation (8.8) into equation (8.5), we can find R(n) to be related to the influencing variables.

In order to account for P, the maximum production rate in the relationships, equation (8.2) is rewritten thus,

$$R(n) = P - \left(P - \frac{n^b}{t_1}\right) \tag{8.9}$$

In order for R(n) to approach P, Levy then multiplies a damping factor, $e^{-\gamma \cdot n}$, to the bracketed term to obtain,

$$R(n) = P\left\{1 - e^{-\gamma n}\right\} + \left(n^b \middle/ t_1\right) \cdot e^{-\gamma n} \tag{8.10}$$

When n is small, equation (8.10) is approximated by equation (8.9), since $\left\{1 - e^{-\gamma m}\right\}$ is close to zero. As n increases $\left\{\dfrac{n^b}{t_1} \cdot e^{-\gamma m}\right\} \Rightarrow 0$, so that R(n) can be approximated by equation (8.5) with a = 0.

Selecting the variable factors

Levy goes into some details to help choose the variables. These he classed in four groups:

(One) Variables associated with planned or induced learning.

These cover all the preplanning processes, the building of prototypes, identification of potential problem spots, etc. These activities help to improve the *initial* efficiency of the process (i.e., improves the value of 'a' in equations (8.5) and (8.6), but is likely to be *inversely related* to the rate of learning, μ.

(Two) Ongoing learning factors.

These may include redesigns of the product (to assist production), re-specification of raw materials, time studies, etc. All factors designed to improve the flow of materials through the plant. The greater the amount of these actions taken, the greater the rate of learning, μ.

(Three) Personnel Factors.

Included here are employee selection, previous experience, incentive plans, and so on. The greater the attention given to employee selection, the greater the initial efficiency and the lower the rate of learning; while incentives would enhance the firm's rate of learning.

(Four) Autonomous Learning.

This includes in-house training; on-the-job learning; learning to take short-cuts; method improvements; learning to cooperate with co-employees; correcting errors; making rapid adjustments, etc.

All these factors tend to enhance the rate of learning.

Applications

1st. **Study of a new printing press:** Manning a new 2-color offset press using teams of three persons working three shifts each day. Data on output from each group was available. Variables used in Levy's analysis are shown in Table 8.4.

Table 8.4: Variables Identified for Rate of Adaptation on the New 2-Color Offset Press (based on Levy, 1965 **)

Item	Variable	Units	Symbol	Group Leader		
				1	2	3
1	Scheduling	None	y_1	117	123	131
2	Training	Days	y_2	14	14	20
3	Printing expense	Years	y_3	16	13	8
4	Total work expense	Years	y_4	4	3	2
5	Shift	None	y_5	$y_5=1$　　　1st shift　$y_5=2$　　　2nd shift		
6	Make-Ready* (M/R)	None	y_6	$y_6=1$ with M/R　$y_6=2$ with *no* M/R		

*At times, one shift 'makes-ready' for the next shift to produce.
**Reprinted by permission, The Institute for Management Science (currently INFORMS), 901 Elkridge Landing Road, Suite 400, Linthicum, MD 21090-2909, USA.

The regression resulting in items 1,3,5 and 6 is significant (0.10 level) with the $R^2 = 0.86$. Surprisingly, item 2, the formal training, did not turn out to be a significant factor. Items 5 and 6 appear to have the largest impact.

2nd. Typesetting new telephone books. Table 8.4 covers the variables used for investigating the adaptation learning rates for system typesetters (relevant data is included in the article).

The regression resulting in items 2, 3, 4 and 6 is significant (0.10 level). Great emphasis is placed on item 4, which justified the emphasis the company placed in its employee selection procedure. Women seemed to appear to be less efficient than men, but there could be other factors at play besides the sex.

Table 8.5: Variables for Adaptation Rate Investigation for Typesetting New Telephone Books (based on Levy, 1965).

Item	Variable	Units	Symbol
1	Education	Years	y_1
2	Training	Days	y_2
3	In-plant Typesetting Expense	Years	y_3
4	External (previous) Typesetting Expense	Years	y_4
5	Total Work Expense	Years	y_5
6	Sex	None	$y_6 \begin{cases} 1 & \text{male} \\ 2 & \text{female} \end{cases}$

3rd. Levy gives additional applications which includes 'budgeting'; evaluation of formal training programs; and 'equipment replacement'.
Without doubt, the technique here depends very much on past data being available. The intention is to learn from the analyses in order to enhance particular factors for future efficiencies in the plant.

8.2.2 Denis Towill (and co-authors): Industrial dynamics family of learning curve models. Denis Towill has been associated with learning curve theory for almost 30 years and has published extensively in the professional literature (see also Towill, 1989, 1990; Naim and Towill, 1990; Sriyanandad and Towill, 1973; and Towill, 1982). The early Towill model for individual learning curves was presented in Chapter 3, Section 3.6. Towill's current work is directed towards start-up operations with particular emphasis (but not exclusively) in the general area of "Advanced Manufacturing Technology (ATM)".

As Towill's work proceeds, we see that his early model, presented in Section 3.6, becomes a part of an increasing "Industrial Dynamics Family of Learning Curve Models", which describes a graded set (by increased mathematical complexity) of learning curves, as illustrated in Figure 8.3. The intention is to fit the models in the family to real data, until a satisfactory selection is found (having an acceptable R^2 value).

The model discussed in Section 3.6, called the "Time Constant Model", appears in third place from the start (the simplest model). A paper by Naim and Towill (1993) goes into the development and application of the "S" model. In this particular paper, the family contained five learning curves, the sixth learning curve, "the Delayed Time Constant model", was included in the family of learning curves published by Towill and Cherrington (1994).

The Towill papers give ample evidence of their applicability using both real and simulated data. When the fit is good, the predictions are also good. However, there lies the problem. With increased learning curve complexity, more parameters need to be evaluated. Even the three parameters needed for the simpler "Time Constant Model", gave my research student much difficulty (Rosenwasser), and this included correspondence with Bevis (one of the authors). After several 'trial-and-error' attempts, the 'best' learning parameters for our data was obtained, but in the ensuing comparative tests (power model, Bevis/Towill, Minter), we found that the Power Model produced superior results to the other two. Towill, himself, is fully aware of this problem (Towill and Cherrington, 1994).

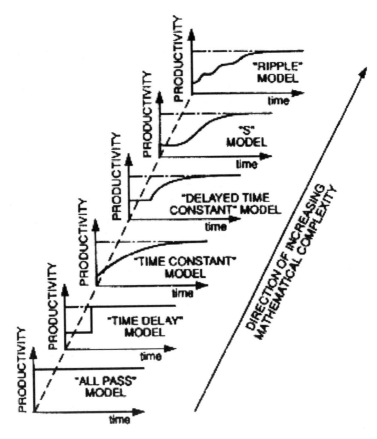

Figure 8.3: Family of transfer-function-based learning curve models arranged in order of increasing complexity

The 'Industrial Dynamics Family of Learning Curve Models' appears attractive and some readers may be tempted to proceed with its method and applications. The mathematical derivations for at least the latter two curves (end of the 'complexity' chain) is beyond the scope of this book and readers are encouraged to go to the source references for help in both understanding the derivations and in applying them in practice.

8.3 Learning at the Firm Level and Econometric Models

It never ceases to amaze this writer, how learning curve models were developed almost independently by both the industrial engineers and the economists. The former group had focussed mainly on individual learning, supporting their work through controlled laboratory experiments, while the economists had concentrated on "cost data" usually only available at the firm level. Both groups have done theoretical studies, as well as practical investigations. The work of the economists tends to be very critical on statistical aspects of their analyses; they have tried to capture important characteristics, such as knowledge acquisition, which normally would be ignored by the engineers (see Dasgupta and Stiglitz, 1988; Spence, 1981; and Fudenberg and Tirole, 1983).

The economists also use terms generally unknown to the engineers. For example, the engineers "learning constant (b)" is referred to as the "Learning Elasticity" - same meaning! They also use the expression "learning by doing" to refer to work actually performed in manufacturing; there is nothing equivalent in the work by engineers and ergonomists.

A distinction is made between learning at the "firm level" and learning at the "industrial level". The limited studies on the former derives from the proprietary nature of cost data belonging to the firm, whereas extracting learning characteristics of the firm from industrial data presents serious assumptions to be made (see Lieberman, 1982).

Applications at the firm level go back many years - in fact, right on the heels of the pioneering work by Wright (1936). Alchian (1963) worked on airframe production; Searle and Goody (1945) analyzed shipbuilding programs during WWII - both studies yielded learning slopes of approximately 80% (which also corresponds to Wright's investigations). Hirsch (1952) analyzed the production of machinery, obtaining a learning slope of 82.5% (which corresponds very nicely with the work by Nadler and Smith (1965), done on typical machining processes).

During the 70's, the learning curve became a familiar concept to corporate managers. Many were influenced by the work of the Boston

Consulting Group (BCG) (1970), who proposed the use of learning curves as a planning and decision-making tool for management, claiming that the learning curves also account for capital, marketing and administrative costs. While the learning curve is an outcome of labor learning, process improvement and product standardization, BCG makes no attempt to decompose the learning curve into these factors. The BCG model is given as follows:

$$c(t) = c(1) \, v \, (t)^{\beta} \qquad\qquad\qquad (8.11)$$

where, $c(t)$ is the real unit variable cost at time t
$\quad$ $c(1)$ is the real unit variable cost at time $t = 1$
$\quad$ v and β are elasticities for the cost of various resources and accumulated volume, respectively.

Webbink (1977) made the first application of learning curve in the semiconductor industry, finding these curves applicable for virtually all products. Grubber (1992) also investigates this area, but also includes product innovations/improvements in his analysis. Not surprisingly, he finds learning slopes in the order of 72%. Harvey and Towill (1981) have also reported the application of learning curves to the steel, paper, automobiles, and chemical industries.

Wormer (1984) developed a "Lagged Model" which he claims suits many cost applications at the firm level. This work is followed by other papers, including one on an integration of the economists' 'cost function' and the engineers' 'learning curve', which could be of interest to engineers (see Gulledge and Wormer, 1990; Dorroh et al., 1986; and Womer and Patterson, 1983). Readers may like to refer to a later work by Wormer and colleagues, Gulledge, Tarimcilar and Wormer (1997) which deals with estimation problems in learning curves. McDonald (1987) produced a new learning curve, which he claimed to be superior to Wormer's model. Heineke (1986) also considered typical estimation errors, which if ignored, can lead to serious errors. This work too provides an interesting insight into the area of econometric issues.

8.4 Learning in Aggregate and Capacity Planning

Aggregate planning deals with the effective allocation of resources (manpower, machines and investments) in the future according to expected forecasts of future activities, with the intention of smoothing workload and minimizing labor and inventory costs.

Until about 30 years ago, researchers in this area had ignored the effects of learning, but with the knowledge that long-cycle repetitive tasks (as in aircraft) experienced shortened cycle times as the task was repeated, led to the inclusion of learning into the aggregate planning models. See, for example, Ebert, 1972, 1976; Khoshnevis and Wolfe, 1983a, 1983b, 1986; Khoshnevis et al., 1982; Behnezhad and Khoshnevis, 1988; and Manivannan and Hong, 1991.

Aggregate planning as a subject is outside the scope of this book and the learning phenomenon is simply an 'add on' to the model; consequently, the reader is directed to a paper by Kroll and Kumar (1989), who have made an excellent survey of this topic - with and without learning.

It is emphasized that there are many books and literally several hundreds of scientific articles written on this subject. It is a favorite topic for university graduate theses.

However, we have yet to see *one* real industrial application!

Apparently, company managers are reluctant to trust major decisions on inventory levels, workforce size and production levels on some mathematical foundations written by university researchers who have had little or no industrial experience.

A more recent article on 'Rough Cut Capacity Planning' (Smunt, 1986, 1996) deals with a similar problem - but applied to the MPS (Master Production Schedule), usually developed in conjunction with MRP (Material Resource Planning) models. This paper is very much 'down to earth', but still shows no evidence supporting actual applications (it is unlikely to be applied until some of the 'strong' learning assumptions are modified). Nevertheless, Smunt's work has some practical possibilities. See also papers by Pratsini et al., 1993; Smunt, 1986b; and Smunt and Morton, 1985.

8.5 Learning in Artificial Intelligence

There are literally hundreds of papers written in this area. They all talk of "Learning Curves", but the application is totally outside our own domain; concentrating in general on machine intelligence or artificial intelligence.

This area is mentioned here, since a library search on "learning curves" unearthed hundreds of references of this kind. Included are four references picked at random, just to give the reader some idea of the paper titles (that mention 'learning curves'). The articles have not been analyzed and are randomly chosen for inclusion (Vezu and Kashima, 1996; Schuurmans, 1997; and Golea and Marchand, 1993).

8.6 The Impact on Quality and JIT

In this section, we are asking a specific question: "Does 'quality' have any impact (or, influence) on learning curves?".

Until now, we have seen that learning curves reduces labor hours, or unit costs, as experience is gained. We also know that the rate of improvement in unit costs (or, labor hours) decreases as experience is accumulated. However, during the last two decades, great effort has been expended on raising productivity via the process of continuous improvement (TQM, JIT and KAIZEN (Imai, 1989)), and without a doubt, these efforts result in improvements in productivity and in the lowering of unit costs and labor hours. Would it, therefore, be correct to assume that 'improved quality *accelerates* the learning curve process'?

This may be 'begging the question', since our usual assumptions are that learning curve improvements *are* due to all kinds of factors, including method improvements, quality improvements, incentives apart from the acquisition of skills. However, the quality link has not been overtly identified as an 'external' factor. Much has been written about continuous improvement. The earliest reference must be Shewhart (1939), whose PDCA (Plan-Do-Check-Act) was later picked up and popularized by Deming (1986). The PDCA cycle goes as follows:

P - Plan: What would be the most important process (or issue) to tackle? What would one like the process to do? What changes must one make in order to achieve the objectives? Plan for the change, define the measures and test.

D - Do: Carry out the changes as planned and measure.

C - Check: Check the measurements against objectives. Verify that the objectives are being met.

A - Act: What did we learn from the results? In what way can we act to make the next change in the process?

The PDCA cycle is repetitive, learning from the previous cycle.

Since then, the "7" tools in TQM have been popularized (e.g., see Ishikawa, 1972) and so has KAIZEN (see Imai, 1989); all those purporting to support the notion of 'continuous improvement'.

An excellent article by Dance and Jarvis (1992) discusses in some detail the application of 'continuous improvement' principles in the semiconductor industry. Other researchers, though, have investigated the link between quality and learning, where quality is measured either as the amount of acceptable products produced, or the amount of defective material produced in the firm. Both approaches will be discussed.

8.6.1 Continuous improvements in the manufacture of semiconductors.
Semiconductors, characterized by their "yields" and 'yield models', are widely discussed in the literature (see Stapper, 1989). Yield models relate integrated circuit yield to circuit design and are subject to defects that have some probability of causing failures.

Process improvement cycles accelerate the learning curve. The specific yield model used is not so important for the improvement cycle - only the methodology of continuous improvement itself.

Dance and Jarvis (1992) describe an application of continuous improvement to semiconductor manufacture. They say that the topic has been discussed by previous researchers.

The following is quoted from their paper[*]:

"Defect-limited yield estimates for a lot may be compared to zone statistics using statistical process control methods, driving the process improvement cycle. Information can be fed back to prior processing to improve process, methods, and equipment. Information can also be fed forward to later zones for quality control and scheduling. Inspection, analysis, and feedback are repeated for each process zone.

After processing is finished, electrical yield is compared with model yield forecasts. Analyzing these differences drives another improvement cycle. If failure analysis shows the presence of nonrandom defect related problems then information is fed back to the process. Otherwise yield model parameters are updated for continuous improvement.

Implementing yield models in the improvement cycle has made it possible for process engineers to quantify their own process sector's influence on [electrical] test yield. They no longer have to wait months to get actual final test results to ensure that the changes they made worked. Thus an improvement cycle using yield models accelerates learning curve progress."

Dance and Jarvis (1992) then proceed in describing some of the improvements (which are omitted because they deal with the technology of semiconductors). However, the conclusion of their study clearly shows the reduction of the average time between the 'lot start' and the 'feedback test results' (which we call the "production lead time"). Figure 8.4 illustrates the relationship between 'lead time' and calendar time (in months) as the continuous improvement process is applied, and a significant negative slope

---

[*] Reprinted by permission, © 1992 *IEEE*.

indicates a sharp reduction in lead times, which can be translated into increased output.

This paper is a fine illustration of the link between 'quality' and the 'learning curve', which should stimulate many firms to apply in their own respective environments.

Median Length

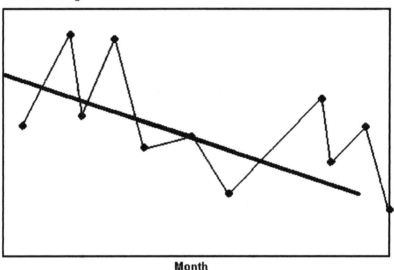

Month

Figure 8.4: Reduction in lead times over time

8.6.2 The affect of quality factors on learning. This section reports on an innovative research on the section topic authored by Li and Rajagopalan (1997, 1998) and parts of their paper are included here.

> "The studies of Japanese manufacturing firms by Garvin (1988) and Abernathy et al. (1981) suggest that quality-related managerial activities may be an important factor in explaining the significant difference in learning rates across firms. When defect levels are high, a firm may devote additional effort to investigate the cause of defects and this leads to additional knowledge of the process which, in turn, increases both quality and productivity. Thus, the defective output may be seen as a trigger or surrogate for the improvement efforts which result in the induced learning effects discussed by Dutton and Thomas (1984)."

> "Induced learning represents the result of deliberate actions by management, workers, and engineers to improve the quality and efficiency of the production process. Examples of induced-

learning activities are: process improvement projects, defect prevention efforts and quality circles."

"In a similar vein, Fine (1986) argues that "firms choosing to produce high quality products will learn faster or go down a steeper experience curve than firms producing lower quality products (where quality is defined as degree of conformance to design specification, not the quality of design)". Fine (1986) presents a quality-based learning curve model where experience gained over time is a function of the cumulative number of good units produced rather than cumulative total production. Extending this line of research, Kini (1994) also develops a quality-based learning curve model and shows that quality increases over time if defective units have greater impact than good units on learning and vice versa."

"Adler and Clark (1991) (discussed in Section 8.1), uses the induced and autonomous learning framework to investigate the impact of labor training and engineering changes on learning, although the quality variable is not considered. Rather than consider the impact of quality on learning, Badiru (1995) examined the effect of learning on product quality. Schmenner and Cook (1985), though not focusing on learning effects, found that the most productive plants were also those that paid the most attention to quality in a study of productivity differences at 95 large North Carolina plants."

Li and Rajagopalan describe four learning curve models which are used in their research (it is sufficient to describe these rather than to also express them mathematically).

Model 1 (Berndt, 1991)
This is the simplest model expressing the direct unit labor hours as a function of the unit labor hour at time "0", the cumulative output and a term accounting for the economies of scale.

Model 2 (Fine, 1986)
Similar in form to Model 1, except that the cumulative output refers *only* to the production of good products, i.e., the total output is multiplied by the proportion of products that are non-defective.

Model 3 (Li and Rajagopalan, 1997)
Similar in form to Models 1 and 2, except that the cumulative output refers *only* to the production of defective products, i.e., the total output is multiplied by the properties of products that are defective!

Model 4 (Li and Rajagopalan, 1997)

Similar in form to the others, except that the cumulative output is split between the proportion of 'good' and 'defective' products. All other terms are the same as used in the other models.

The three hypotheses tested by Li and Rajagopalan were as follows:

H1:   That the cumulative output of *good* products is significant in explaining the learning curve benefits (Model 2).

H2:   That the cumulative output of *defective* products is significant in explaining the learning curve benefits (Model 3).

H3:   That the cumulative output of good *and* defective products does not equally explain the learning curve benefits, and in particular, the cumulative output of defects better explains the learning curve benefits than the cumulative output of good units.

Data was collected over a period of three years from two leading manufacturers: Firm A and Firm B.

Firm A produced car tire thread by extrusion and pressing processes, while Firm B manufactured a variety of medical instruments used in critical-care units in hospitals to monitor vital signs (quality here is a critical factor). Both firms made serious efforts to reduce defects (or, improve quality) via 'induced' improvements, i.e., training in use of quality tools, regular preventive maintenance, SPC (Statistical Process Control), job rotation, group meetings to discuss quality problems, and so on. Furthermore, there were no major changes in production processes, or, technologies during the three years of data collecting.

Results of the tests are summarized in Table 8.6.

Table 8.6: Results of the Tests (based on Li and Rajagopalan, 1997a)

Test	Learning Slopes Firm A	Firm B
Model 1: Explain L by total output	92.8%	95.4%
Model 2: Explain L by good output	92.8%	95.4%
Model 3: Explain L by defective output	81%	91%
Model 4: Explain L by either good and/or defective output	Defective items better explain the learning process than does the production of good items.	

In all tests, the 'economies of scale' factor was found to be insignificant in both firms. "The finding that defective units help explain learning curve benefits is consistent with anecdotal evidence from Japanese manufacturing. For example, Hayes (1981) and Schonberger (1982) point out that Japanese firms treat each defect as a "treasure" for uncovering the sources of imperfections in the production process and for helping workers and engineers in learning to improve the production process. The finding is also consistent with state-of-the-art quality management practices in the U.S. Motorola, for example, makes use of each defective unit to find the "bugs" in the process and then improve the process (Omachonu and Ross, 1994)."

Without a doubt, the generation of defective items stimulates the firms to take countermeasures to eliminate the problem and in the process, often find simpler and more efficient ways for producing the items. It would imply that if all goes well (quality-wise) then perhaps these processes would be left untouched, while attention is focussed on the problem areas, which become the ones to reap the potential learning benefits.

Many of the improvement processes discussed in this chapter are also relevant to Learning Organization, the topic of the next chapter.

9 LEARNING ORGANIZATIONS

It is fair to say that "Learning Organizations" is one of the hottest topics in today's business world. "The rate and effectiveness of Learning Organizations may become the only sustainable competitive advantage, especially in the knowledge-intensive industries" (Stata, 1989).

This is typical of the strong statements expressed by adherents of Learning Organizations. There is, therefore, a need to explore its meaning, structure and principles, and to find out in what way these differ (if at all) from 'Organizational Learning'? Is there any connection between Individual Learning and Learning Organizations?

These factors and more will be considered in this chapter under the following headings:

9.1 Does 'Learning Organization' differ from 'Organizational Learning'?
9.2 A Primer on Learning Organizations
9.3 Linking individual and plant learning to Learning Organizations
9.4 Teams and Team Operation
9.5 Developing into a Learning Organization

9.1 Does Learning Organization Differ from Organizational Learning?

If we turn to the experts, one could conclude that the two titles refer to identical concepts and are used interchangeably. For example, Peter Senge uses the term 'Learning Organization' throughout most of his book, 'The Fifth Discipline'. But he has also used 'Organizational Learning' on occasion, and both terms appear separately in the "Subject Index".

Personally, I would prefer that their meanings be different. Organizational Learning should refer to the vast knowledge developed and implemented for improving traditional organizations. These organizations have developed visions, purposes and core values, and many have been, and still are, highly successful enterprises. Their operating approach is essentially one of pragmatism – do today what works; and whatever works today, may not work tomorrow. This view is reinforced by current management's emphasis on solving problems.

The term 'Organizational Learning' has been around for at least twenty years; Argyis and Schon authored a book entitled "Organizational Learning" in 1978. There are others too: Duncan and Weiss, 1976; Hedberg, 1981; and Fiol and Lyles, 1985. Senge adds that traditional organizations are designed to keep people comfortable, to inhibit taking risks, and to make changes by reacting to events. But such organizations have invested much energy in making improvements in company performance using TQM, Reengineering, external consultants and a host of other techniques concerned with the continuous improvement process.

Organizational Learning would, therefore, be involved in the improvement process in virtually all the industrial studies discussed in the early chapters of this book, especially at the plant level and on learning associated with 'products' development. The term 'Organizational Learning' was used for years in association by writers such as: DeJong, 1957; Kilbridge, 1962; Kottler, 1964; Levy, 1965; and Globerson, 1980, 1993.

On the other hand, the Learning Organization is an entirely new ball game with its emphasis on systemic thinking and its five learning disciplines: systems thinking, personal mastery, mental models, building shared visions and team building. These terms and more, will be discussed in some detail in the next section.

9.2 A Primer on Learning Organizations

"A Learning Organization is one that is continually expanding its capacity to create its future" (Senge, 1990).

A dictionary definition of a Learning Organization does not give too much away since it would simply be defined as an 'organization' (take any definition) that learns. As a consequence, we need to define Learning Organizations in terms of its characteristics. For example: "A 'Learning Organization' is skilled at creating, acquiring and transferring knowledge, and at modifying its behavior to reflect new knowledge and skills" (Garvin, 1993). Or, "Organizational knowledge is embodied in its standard operating procedures, routines, common perception of past events, common goals, shared assumptions, architecture and strategic behavior" (Levitt and March, 1988). And: "....Capacity or processes within an organization to maintain or improve performance based on experience" (Nevis et al., 1995).

These definitions help in our understanding, but still do not capture the scope and depth of the understanding behind Learning Organizations.

We get a much clearer picture of Learning Organizations from Peter Senge himself. In his profound book on Learning Organizations, Senge

(1990) refers to "… system thinking as the fifth discipline because it is the conceptual cornerstone that underlies all of the five learning disciplines …".

Senge goes on to describe the five learning disciplines as:

- **Systems Thinking**: " is a discipline for seeing wholes. It is a framework for seeing interrelationships rather than things, for seeing patterns of change rather than the 'static' snapshots." "Systems thinking is needed more than ever before because we're becoming overwhelmed by complexity." "We create far more information than anyone can absorb." "Some issues can be understood only by looking how major functions such as manufacturing, marketing, and research interact; but there are other issues where critical systemic forces arise within a given functional area; and others where the dynamics of an entire industry must be considered."

- **Personal Mastery**: "The discipline of personal growth and learning." "It is a process and a lifelong discipline. People with a high level of personal mastery are acutely aware of their ignorance, their incompetence, their growth areas. And they are deeply self-confident."

- **Mental Models**: "… deeply held internal images of how the world works, images that limit us to familiar thinking and acting." "Contemporary research shows that most of our mental models are systematically flawed. They miss critical feedback relationships, misjudge time delays, and often focus on variables visible or salient, not necessarily high leverage."

- **Shared Vision**: is a concept that is truly shared among people. "Few forces in human affairs are as powerful as shared vision." "Shared vision emerges from personal vision." "Shared vision is vital for learning organizations because it provides the focus and energy for learning." "Visions are exhilarating. They create the spark, the excitement that lifts an organization out of the mundane."

- **Team Learning**: "Team learning is the process of aligning and developing the capacity of a team to create the results its members truly desire." "It builds on personal mastery, mental models on the discipline of shared vision…" "Team learning is vital because teams, not individuals, are the fundamental learning units in modern organizations. This is so because all the important decisions are now made in teams, either directly or through the need for teams to translate individual decisions into action."

Kofman and Senge include an essay on "Communities of Commitment: The Heart of Learning Organizations" in Chawla and Renesch's book, *Learning Organizations* (1995), where they state, "… when we speak of a

Learning Organization, we are taking a stand for a vision for creating a type of organization we would truly like to work within and which can thrive in a world of increasing interdependency and change." The authors take a hard look at our modern world behavior, emphasizing the areas of cultural dysfunction, fragmentation and competition. They go on to state that "Learning Organizations must be grounded on three foundations:

(One)   A culture based on transcendent human values of love, wonder, humility and compassion.

(Two)   A set of practices for generative conversation and coordinated action.

(Three)   A capacity to see and work with the flow of life as a system."

What is troublesome is that most of the terms used in the three foundations are 'unscientific' in the sense they cannot be properly defined let alone scientifically measured (the same is also true for much of the five learning disciplines).

So how should we interpret these statements?

In their article, Kofman and Senge go deep into the meaning of life and it is from that point they come to the conclusion "... we must create alternative ways of working and living together. We need to invent a new, more meaningful model for business, education, health care and government and family." We thus can understand the gist of their three formulations without being too critical of its content.

But the problem of interpretation remains!

Handy, in an essay entitled "Managing the Dream" in Chawla and Renesch (1995), talks about the process of change in business – especially those exposed to discontinuities. "When twenty or more countries around the world can produce Chardonnay wine to rival the French, nothing seems sacred anymore. It's no wonder the average shelf life of corporations is only forty years, before they are swallowed, die or merge." Handy goes on to argue that the only way out is to acquire new learning habits – for both individuals and organizations. It's a habit that will alter many of the current assumptions about management, and leads to the creation of the "Learning Organization"

Handy states that the Learning Organization is built on an assumption of **"competence"** which is supported by four characteristics: **curiosity, forgiveness, trust** and **togetherness**. The assumption of competence means that individuals operate at the limit of their respective capabilities with the minimum of supervision. Handy argues that organizations today operate on an assumption of **incompetence** whose characteristics are controls and directives, rules and procedures, layers of management and POWER.

Handy is somewhat extreme in his statement, but he is emphasizing a point. Leadership styles in industry are not 'black' and 'white'. Today's organizations range from Handy's 'incompetence' model to ones that are very close in structure and operation to Senge's Learning Organization (and who don't call themselves learning organizations!).

Handy goes on to explain that '**competence**' alone is not enough to foster the learning habit. It must be accompanied by curiosity, which in turn, raises questions, to which the truly curious must search out the correct answers – sometimes through experimentation.

This process is encouraged by the learning organization. Strictly speaking, when properly performed, there are no "failed" experiments, since we learn from these 'unsuccessful' investigations. Therefore, **forgiveness** is a necessity for unsuccessful experimental work. We learn from our failures and persistence in experimentation could eventually result in breakthroughs.

Curiosity and forgiveness must be accompanied by **trust**, since we must trust the competent person to experiment in the first place (within the terms of his/her competence). We reduce the risks in experimentations by developing a sense of **togetherness,** since many problems faced by business today should be **shared** among the key members of the organization who collaborate and learn from each other.

Despite the presence of trust and togetherness, Handy argues that the Learning Organization may not be a comfortable place for its leaders since much of the power resides outside top management's jurisdiction. Imposed authority no longer works and must be earned from those over whom it is exercised. The organization is held together by shared beliefs and values by people committed to each other and to common goals.

Handy goes on to describe the learning process as a turning wheel comprised of four quadrants,: (in order) Questions, Ideas, Tests and Reflections. This concept is remarkably similar to the Shewhart/Deming PDCA (Plan-Do-Check-Act) cycle, used so effectively in the continuous improvement process. 'Questions' may be triggered by problems or needs that require solutions. The questions then trigger the search for 'Ideas', which are rigorously tested to see if they work. These tested ideas must then be 'Reflected' upon until the best solutions are identified.

Keeping the wheel in motion requires strong leadership and a belief in the potential for excellence. Handy then goes on to describe the different aspects of Learning Organizations.

Garvin (1998) summarizes Organizational Learning as involving three overlapping stages:

a) The cognitive stage, in which organizational members are exposed to new ideas, expend their knowledge and begin to think differently.
b) The behavioral stage, at which point employees begin to internalize new insights and alter their behavior.
c) Performance Improvement: measurable improvements in organizational results which emanate from stages a) and b).

The above stages give the appearance of capturing the essential practical aspects of the Learning Organization – but it lacks the conceptual depth that was covered earlier.

The definitions and characteristics of Learning Organizations given thus far are very idealistic. For example, Handy states, "There is little wonder at the fact that we have no example organizations that have got it all right."; and referring to the work done at MIT's Learning Center, Peter Senge writes, "Out of 13-14 major projects, we have had spectacular success and we have stopped about half the projects within two or three months."

The projects Senge refers to are the work done by specific teams from participating organizations – teams, whose membership numbers would represent an insignificant fraction of the total manpower of each respective organization. Are they really representative of the style of work done in their respective organizations?

Senge (in Ayas and Foppin, 1996) reported the result of one project "... developing the next generation of the Lincoln Continental", and claimed a saving of $65,000,000 in development expenses – a no mean achievement.

It is unavoidable to arrive at the conclusion that the description of "Learning Organizations" refers to some utopian state that few organizations may achieve in totality. Senge mentions several companies that may fall in this category: Kyocera (Japan), Hanover (US), Analog Devices (US), Herman Miller (US), Shell (The Netherlands) and others. Semco (Brazil), may also fall in this category (Semler, 1993). This doesn't prevent individual teams in other companies from operating within this framework. But unless the process is intentionally spread throughout the organization, the success of these teams is likely to be lost as the team stops functioning. This can occur at the completion of a project, or else, as key members of the team (because of their success) get relocated to more responsible positions – either, within the organization, or, with some outside firm.

What has been described as Learning Organizations must involve the "Change Process", and the Change Process has been with us for over half this century. Does every application of a 'Change Process' constitute an element of the Learning Organization team? One may be inclined to answer "Yes" to this question, even for simple 'methods improvement' teams, since these must consider such questions as, "Is the action necessary?" and "How else can this action be achieved?" – questions, which, if properly considered, require a concentrated analysis of what is being considered, and on whether its existence is at all necessary. But do these teams work with a shared vision? I think not!

Section 9.3 will consider the many 'formal' team processes that have come into existence, all of whom may be considered as 'initial' elements of Learning Organization teams – though none have made any pretense of claiming to be a part of the Learning Organization. On the other hand, the projects Senge (in Ayas and Foppin, 1996) refer to, *are* elements of the Learning Organization. So too are other applications that have made it to the literature in recent years, and are discussed in the next paragraphs.

9.2.1 Learning Organization applications. In an applications paper on product development at the Whirlpool Corp., Duarte and Snyder (1997) attempt to address all elements of the learning organization, which includes:

- gathering information
- interpreting data
- sharing lessons learned
- storing of critical knowledge

The authors used a series of "best practices" exchange conferences, and included as participants, Whirlpool employees from the US and from its companies throughout the world (and in some cases, specialists from outside the Whirlpool Corporation were also invited to participate). This provided an excellent source for information gathering, for shared 'key learnings (practical knowledge and conclusions) and for networking.

Durate and Snyder's product development model was based on Huber's (1991) learning organization model and on Wheelwright and Clark's (1992), model for learning after product development (see also Hayes and Wheelwright, 1984). Their model included the four main elements discussed earlier, and detailed below:

1) Information gathering and filtering success factors in product development from both internal Whirlpool sources and external sources. It included factors relating to both success and failures; on best practices; real time exchange with team members; academic and practical literature.

2) Methods for analyzing the information, including a formal framework for analyzing key learnings from product development teams.

3) Methods for storing information about best practices and key learnings, including training courses, documentation, using people and culture.

4) Ways to distribute information on a JIT basis to people and product development teams. Methods included team mentors, training, interventions, documentation and handbooks.

One can view the work by Durate and Snyder within the context defined by Handy – with the accent on 'competence'. Full trust was placed on the two authors, specialists in 'Organization, Leadership and Human Resource Development', who in turn, studied, experimented and tested various ideas for meeting their objectives. We gather from the article that excellent results were obtained by the various product performance teams at Whirlpool. We must accept that the concept of 'shared vision' was a given among all participants.

Globerson and Ellis (1995) and Ellis and Globerson (1996) describe the state of Learning Organization being practiced by 15 companies dealing with projects (though, none claimed to be Learning Organizations). Forty-four project managers, working in the 'High Tech' area, were asked to fill out an "Organizational Learning Diagnostic Tool" which was subsequently analyzed.

The Diagnostic Tool contained 34 questions pertaining to three aspects of Learning Organizations:

(One)   Knowledge creation: how information was gathered in order to improve knowledge.

(Two)   Information diffusion: how the organization facilitates the conversion, or, elaboration process.

(Three) Information storage and retrieval: attributed importance of organizational knowledge.

Respondents were asked to rate their agreement on a 7-point scale ('1' means low agreement and '7' means high agreement) over the 34 questions.

What was more relevant from the analysis were the two tables covering questions that had a high level of agreement among all participants (score > 5.4) and those that had a low level of agreement (score < 2.5). The two tables are shown below as Tables 9.1 and 9.2 respectively.

Table 9.1: Items that drew a high level of agreement among all participants (score >5.4) (from Globerson and Ellis, 1995)

1	Similar problems repeat themselves from one project to another
2	Information gathering and analysis is an inherent part of the job
3	Each employee is expected to pass on data to other employees
4	I have information concerning the areas of expertise of individuals in my company
5	I am very often approached by other people who require professional advice
6	In my opinion the subject (of the questionnaire) is important

Table 9.2: Items that drew a low level of agreement among all participants (score < 2.5) (from Globerson and Ellis, 1995).

1	A high proportion of my work is guided by written routines
2	My company maintains a coding system to group work according to similarity
3	I very often participate in outside seminars
4	My organization is very permissive about the use of a 'trial and error' approach
5	My organization assigns a relatively large budget for learning

The companies in Globerson and Ellis's study had no expressed ambitions of being Learning Organizations. Yet, many items important to Learning Organizations were practiced by those 15 organizations, while other aspects important to Learning Organizations (such as, shared vision, systemic thinking, and others) were omitted from the Diagnostic Tool.

While the questionnaire provides some interesting results, it does not sufficiently define all elements of the Learning Organization as defined by Senge (1990) nor can a simple score on 34 questions replace an in-depth dialogue between the researcher and practitioner. Nevertheless, the diagnostic tool provides plenty of evidence for a Learning Organization practitioner to identify aspects of a company's operating practice that could do with further support and development.

The link between TQM and Learning Organizations should be fairly obvious. As a managerial system for continuous improvement, TQM teams have gradually evolved into self-managed (or, self-directed) work teams, whose mode of operation come pretty close to the work team under a

Learning Organization. Hill (1993) talks of learning organizations for TQM via Quality Circles (QCs) which is a bit unfortunate, since QCs, as a 'bottom-up' approach (as practiced in the West), has a very low chance for survival (a view supported by Senge, 1990), since these contain the seeds for their own destruction (see Dar-El, 1985). Hill (1996) reports a 10-year survey of 28 companies. In that period, only 2 of the original 28 companies still used QCs. However, in Japan, QCs are integrated into a company's TQM approach, and are initiated and supported by top management.

Barrow (1993) talks of adding 'learning agendas' built into a TQM initiative to yield a Learning Organization environment. These aspects will be discussed in Section 9.4.

9.3 Linking Individual and Plant Learning to Learning Organizations

Early in this book (in Chapter 2), we covered the main factors that influence individual learning. Among the factors were, "Continuous Improvement Teams" (Section 2.11), which is the operational basis for TQM. The concept of Continuous Improvement dates much earlier than TQM – Quality Circles, for example, were developed in the mid-50s; KAIZEN, which means 'small improvements' (Imai, 1989), is *the* vehicle used by JIT in the improvement process, and this too, came out in the early 60s.

It was already pointed out that individual learning involves two processes:

One)   The acquisition of skills for performing tasks, and would include simple improvements such as repositioning of tools and products being worked on. Only in laboratory-type controlled experiments, can we be sure that this is the only type of learning that applies, since studies on industrial applications are likely to also include induced changes aimed at improving performances.

Two) Induced improvements, designed to improve performances. These include a variety of tools and techniques, such as, JIT, SPC (Statistical Process Control), continuous improvement teams, experimentation, value analysis, and so on.

The first point, the acquisition of skills appears to be a long way from the Learning Organization methodology. However, this writer argues that it is individuals who comprise the Learning Organization teams, and as such, it is the individuals that need to acquire the skills for operating in teams within the Learning Organization environment. The training of these individuals is done on a more sophisticated level than that for simple tasks, but could be comparable to those being trained for monitoring complex operational and decision-making activities. So the link to individual learning should be clear even at this elementary level.

As to the second point (induced learning), virtually every technique used in this connotation would also be included in the 'bag of tools' used by work improvement teams at all line levels of the organization. Learning Organizations also use 'line level' teams. There are strong similarities between the two – in both cases we have shared beliefs, as well as the ability to redefine the objectives of the exercise (see Haworth, 1996). Redefining objectives is a standard approach used by industrial engineers for over 50 years. For example, Nadler wrote about determining the correct objective level for tackling a problem, in several of his books (1964, 1970, 1981) (see also Nadler and Habino, 1998).

TQM is yet another approach that has strong links to Learning Organizations, and many articles have appeared that discuss this linkage (see for example, Barrow, 1993; Aston et al., 1990; Haworth, 1996; Giunipero and Vogt, 1997; Sexton, 1994; and Magjuka, 1992).

TQM practices in recent years have triggered yet newer approaches to teamwork. It started with talks on employee 'empowerment', but this quickly moved to SMT (Self-Managed Teams), sometimes referred to as SMWT (Self-Managed Work Teams) – though SMT is now the commonly adopted term. SMTs do not necessarily require TQM as a base for its operation, though more often than not, SMTs are imbedded in TQM implementations. SMTs are the closest in operation and character to Learning Organization teams and will be discussed in the next section.

Finally, the literature review on plant learning (Chapter 8) makes it clear that both individual learning (acquisition of skills and 'induced' learning) as well as the use of improvement teams (used under the guise of any of several names (see Section 9.3)) operate in those studies. In many cases the improvement process is extended to incorporate product development, with, or without concurrent engineering methods.

Team activities today, directed towards product development, methods improvement, continuous improvement and the like, in many respects operate as Learning Organization teams would be expected to operate. While their respective CEOs may not even be aware of Learning Organization, they, as a rule, support the independence and contribution of operating teams in solving specific problems.

The following Section 9.4, deals specifically with team activities.

9.4 Teams and Team Operation

In order to survive, organizations are constantly making improvements in strategies, procedures and operations. And improvements mean CHANGE! From the turn of the century until the mid-70's, the major improvement vehicle was Methods Engineering, usually performed by individuals trained in "Time and Motion" (or, "Work Study") analysis. This is not to say that teamwork was not employed for performance improvement during this period. Indeed, by the mid-70's, organized team activities included: Work Simplification (see Morgenson, 1963); Scanlon Plans (Lesieur, 1958; and Moore and Ross, 1974); Rucker Plans (Rucker, 1962); Value Analysis/Value Engineering (Miles, 1961; and Mudge, 1971), JIT/KAIZEN (Scheonberger, 1982; and Imai, 1989) and Quality Circles (Ishikawa, 1972). From the mid-70's onwards, formal team activities sprang up like mushrooms, which included: Concurrent Engineering (Gu and Kusiac, 1993; and Salomone, 1995); QFD – Quality Function Deployment (Akao, 1990; and Cohen, 1995); SHRED COST Plans (Dar-El, 1986); TQM – Total Quality Management (Deming, 1982, 1986); Reengineering (Hammer and Champy, 1993); SMT – Self-Managed Teams (see for example, Magjuka, 1992); and more.

Teams of whatever kind or name, do not suddenly appear 'out of the blue'. Considerable managerial thinking went behind each introduction – especially on the possible impact such systems could have on the organization itself. For example, TQM and Reengineering, each in their time, were radically new concepts, whose introduction would markedly alter organizational behavior. Other teams, such as Concurrent Engineering, VA/VE, QFD, etc., had only localized impact with virtually no influence on managerial behavior.

Virtually all reported teamwork was done in industry and under industrial conditions. Thus, one can never claim a unique casual relationship existing between team-induced changes and the measured outcomes, since many other uncontrollable factors could also operate in parallel, confounding the causes of the measured improvements. But the general attitude should be, "Who Cares?" – especially when we achieve the purpose of the exercise – i.e., the long-term improvement in productivity.

And is this not also the ultimate purpose of Learning Organizations?

Learning Organizations also use teams and team learning for resolving problems and for developing improvements (read, CHANGE!). The framework surrounding these teams is carefully defined, but it functions with individuals who are charged by a shared vision and philosophy for achieving their objectives.

Are the formal teams mentioned, all that different from the Learning Organization team? The next section reviews some of these formal teams, their behavior, characteristics and operation.

9.4.1 Work teams - past and present.

a) Work Simplification appeared in the mid-40s, the brainchild of Morgenson (1963), in which cross-functional teams were created in order to solve specific problems found in daily operations. Already at the time, team members were charged with the responsibility for redefining objectives in case the problem could be eliminated altogether. Work Simplification teams peaked in 60s with excellent results reported by major companies, such as, Proctor and Gamble (see Lehrer, 1957). The method simply died thereafter as other team methods came into being.

b) Scanlon Plans named after Joe Scanlon (see Moore and Ross, 1974) for his proposal for productivity improvements, which was implemented by the Empire Steel and Tin Plate Co. which was close to declaring bankruptcy during the mid-40s. The plan helped the company to recover within the year, and the method become known as the Scanlon Plan. The Scanlon Plan has two components: teamwork wherein proposals for improvements are created, developed and assessed for acceptance, and a reward system based on some 'gain sharing' formula. Unfortunately, over the years, practitioners played down the team activities and the Scanlon Plan became primarily associated with the reward system (which was not a very effective one in the first place!). On the other hand, the structure and operation of the teams were quite innovative and operated in an exceedingly fair manner. Scanlon Plans still exist today – but in a small number of applications.

c) Rucker Plans used an identical method as Scanlon Plans for improving productivity, but the gain sharing rewards method was quite different. Rucker Plans are a propriety system, but to the best of this writer's knowledge, the method is no longer in use.

d) VA/VE – (Value Analysis/Value Engineering) uses a cross-functional team for generating the least cost functional design for a specific product. The approach was developed during the early 60s by Miles (1961), reaching its peak in popularity during the late 60s and early 70s, when DOD contractors were given the opportunity to keep 50% of all savings they could generate by applying VA/VE to any component of the DOD contract. VA/VE members try to update their knowledge on 'best practices' in order to achieve the least cost solution to the product design. Today, VA/VE is still applied, but is more often

integrated with QFD (Quality Function Deployment), which is
discussed later.

e) Quality Circles (QC) were developed in Japan – probably in the mid-
50s, but only became known in the West around the early 70s
(Ishikawa, 1972). QCs originated as a 'bottom-up' movement, but
today, Japanese managers are heavily involved in the creation, training
and operation of QC teams – these being part of their TQM movement
in practice.

After an exciting beginning in the US and Europe during the late
70's, the QC movement virtually died out as an independent
improvement technique. Other types of similarly structured teams are
now a part of most TQM applications, these having been absorbed into
the functional structure of the TQM process itself. One thing is certain,
QCs are unlikely to survive independently of some 'top-down'
technique (such as TQM) which absorbs QCs as a part of its ongoing
operation.

QC teams provide practitioners with the maximum independence for
defining their objectives, alternatives, processes and methods for
tackling their respective problems. In my view, this comes closer to
what is expected of teams operating under Learning Organizations (see
Hill, 1996; and Howorth, 1996).

vi) Concurrent Engineering came into existence around the late 70s and
was used for improving the R&D and Engineering processes within a
company (see Gu and Kusiak, 1993; and Salomone, 1995). Concurrent
engineering utilizes a team comprised of representatives from design,
production, purchasing, logistics, plant engineering, and so on, who
coordinate their skills, knowledge and responsibilities to maintain
updated knowledge on new product development. Information transfer
times disappear and potential design problems are often nipped in the
bud before these get adopted, developed and then revised. The result is
reduced lead times from conceptual design to manufacturing and a
greatly reduced number of design changes that usually occur during
manufacturing.

Embedded in the concurrent engineering process is QFD (Quality
Function Deployment), which is a team approach that takes the
customers' needs into account during the design process. QFD
originated in Japan in the late 70s and is extensively used in the US,
Europe and Israel (see Akao, 1990; and Cohen, 1995). The end result
of a QFD analysis is a set of design specifications for R&D researchers
to consider in their design. VA/VE is logically applied at this stage.

QFD members must update themselves with the latest technology in order to achieve the least cost solutions to their designs.

Within the R&D context, Concurrent Engineering and QFD teams are not only cross-functional, but are very sensitive to marketing and manufacturing issues. Many of the more successful teams would fit very nicely to the needs of a Learning Organization.

vii) <u>SHRED COST</u> (Dar-El, 1986) was introduced as a continuous improvement process as well as a gain sharing reward plan based on CHANGE – "Work Smarter Not Harder" is the guiding motto for employees working under this plan. The improvement process is based on Work Improvement Teams (WITs), aimed at creatively generating solutions to various problems. The improvement process is structured at three levels: the Top Management Team (TMT), the Middle Management Team (MMT) and the Work Improvement Teams (WITs) as illustrated in Figure 9.1. The Top Management Team (TMT) develops the strategy for the company and determines the terms for operating SHRED COST. The Middle Management Team (MMT) is by definition, a cross-functional team; it acts as the conduit for information moving between the TMT and the WITs.

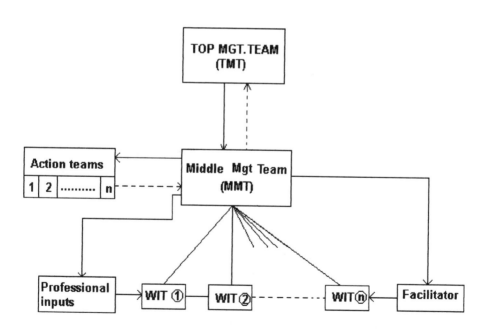

Figure 9.1: The SHRED COST structure

Action teams are utilized for resolving specific problems and which usually also require professional inputs. WIT members are trained on a continuing basis – more often than not, using internal resources. As with TQM, the improvement process acts as the main motivator. The gain sharing rewards simply 'reinforces' the improvement process. It is not recommended to use the gain sharing reward system <u>without</u> the improvement process.

SHRED COST incorporates a total involvement process of the organization as seen in Figure 9.1. In its most effective state, it requires a strong participatory style of management, where employees have great independence in defining problem areas and in searching out and experimenting with possible solutions. Management plays a supportive role in enabling the teams to operate unhindered. It must be stressed that this writer has yet to find an organization operating in a total participatory style – even among Israeli Kibbutz and Trade Union industries (Dar-El, 1993). But without a doubt, some sections of such organizations exhibit strong participatory characteristics.

viii) <u>TQM (Total Quality Management)</u> is both a philosophy and a set of guiding principles designed to create conditions for continuous improvement throughout an organization (Dar-El, 1995). Team members have shared visions, independence and exhibit trust in each other when tackling problems. They are encouraged to experiment when developing solutions. Team members are highly motivated in what they do – even to the extent that financial rewards can be dispensed with altogether (Dar-El, 1992), it being claimed that the mere participation in the creative improvement process provides sufficient motivation without resorting to financial rewards.

TQM provides a natural framework in which to embed organizational learning, since it is designed to encompass *every* aspect of an organization. One of its main channels is through the provision of quality products and services. The ISO-9000 series is now used almost universally as a 'stamp of approval' for quality, but ISO-9000 is also designed for enhancing a TQM environment (see Hoyle, 1998; and Goetsch and Davis, 1998).

Operationally, TQM teams probably come closest to what would be required of teams operating under learning organizations (for example, see Barrow, 1993; Hill, 1995; and Ashton et al., 1990).

Should we conclude that companies totally committed to TQM are equivalent to Learning Organizations? The answer is "NO". So what is missing?

For a start, until a few years ago, TQM was almost totally absorbed in "quality issues". Indeed, even Deming (1986) held the view that an organization need only concentrate on "quality" and business would look after itself! (see also Gitlow and Shelly, 1987). The initial criteria for the Malcolm Baldridge Quality Award was based on similar assumptions. Nothing could be further from reality. Quality is only one ingredient in the productivity equation and devoting energy to it exclusively is a recipe for disaster! For example: Florida Lighting and Power won the coveted Deming Prize in 1989 (the only company outside Japan to do so). A year later, the company lost some $390 million. Under the new CEO, about 95% of its original TQM team were 'retired'. IBM was the Malcolm Baldridge Award winner in 1990. In 1991, IBM's CEO Akers was elected chairman of the National Quality Month. In 1992, IBM ended the year with a net loss of nearly $5 billion. One hundred and fifty thousand employees were laid off or retired. In 1992, Wallace Inc. files Chapter 11 (bankruptcy) after wining the Malcolm Baldridge award the year earlier (see, for example, Rao et al., 1996). And so it goes.

Going bankrupt is certainly not the objective of Learning Organizations, nor the objective of any business enterprise. Yet this is what happened with some companies perusing TQM exclusively focussed on Quality. For a period during the early 90's, TQM was almost a 'dirty' word – reengineering became the new 'savior'. TQM has since redefined itself, keeping the bottom line clearly in focus. Even the Malcolm Baldridge criteria is now heavily biased to economic success factors in its measurement process (see, for example, Rao et al., 1996).

We need to go back to Peter Senge's description of the five learning disciplines at the beginning of this chapter, to see that only parts of the disciplines overlap TQM concepts. These come from personal mastery, shared vision and teams learning. But Learning Organizations only come when all five learning disciplines are present. Nevertheless, several cases are reported where companies working with TQM are trying to broaden their activities into Learning Organizations (Rogers, 1998).

i)   Self-Managed Teams (SMTs) were an outcome of the early work on empowerment raised by TQM practitioners. What gradually evolved were Self-Managing Teams (SMTs) which involved a number of workers who manage themselves in performing a significant unit of work (see Yeatts and Hyten, 1998). In many respects SMTs are an outgrowth of Quality Circles (see Sims and Dean, 1985), but they have gone far beyond QC activities (see Tjosvold, 1991). For example, SMTs are given the responsibility to handle task assignments, scheduling, coordination with other groups, setting goals, performance evaluation and discipline. They often replace the function of supervisors (Donovan, 1986).

SMTs are committed to a common purpose, a set of performance goals and hold themselves to be mutually accountable. Teams should develop strengths in technical expertise, problem-solving, decision-making and in interpersonal skills. They must be created with proper training, shared vision, shared values, shared benefits, trust for each other and supported to some level of risk-taking (see: Sexton, 1994; Mikalachki, 1994; Sagoe, 1994; Ward, 1997; Moravec, 1998; Carroll, 1998; and Tanskanen et al., 1998).

Clearly from the above, the characteristics and functional operations of SMTs come very close to those of Learning Organization teams.

Dramatic productivity improvements and outright savings are the expected outcomes of SMTs as reported in many references (see for example, Moravec et al., 1998; Elmuti, 1996; Spencer, 1995; Hitchcock, 1993; and Klepper et al., 1989).

j) Reengineering entered US industry in the early 90s with a "bang" (Hammer, 1990; Hammer and Champy, 1993). "Keep the walls standing and 'nuke' the rest" was how Hammer introduced reengineering. The method begins with a blank sheet of paper with a team appointed to redefine the best way that work should be accomplished – the accent being on the processes needed for accomplishing the work. The teams are given full autonomy in defining the direction they should take and are encouraged to disregard existing procedures, operating rules and assumptions. They are also encouraged to test out their ideas in order to identify the most effective methods to use.

Without question, this invigorating approach resulted in the redundancy of large numbers of employees, especially of middle managers who joined the ranks of the unemployed in droves. This downsizing phenomenon was an inevitable outcome of reengineering, and was the main reason for the incredibly good financial outcomes with its implementations. But even this situation was too much for US industry, and today, Reengineering is off the popularity list! Even Hammer has admitted his 'error' in ignoring the human factor (see Hammer, 1996).

There is little to be gained in detailing more team activities that have appeared in the literature. Each method has made a contribution, some more than others.

9.4.2 Single-loop vs. double-loop learning in teams. The concept of single-loop vs. double-loop learning was proposed by Argyis (1977), Argyris and Schon (1977), and supported by many others (see Lundberg, 1988; and McKee, 1992). Single-loop learning refers to a ready response built into the organizational memory. For example, the general manager asking for a budget to publicize the firm, but insists that the team secure the services of a specific advertising agency, is an example of single-loop learning.

Double-loop learning is described by Argyris (1991, 1992, 1996), as challenging, exploring and transforming accepted norms of behavior – it is virtually identical to the approach used in reengineering (Hammer and Champy, 1993). For that matter, virtually every type of team discussed in this section should operate in this manner. Indeed, in practice, one would find *very few* examples (if any) of teams that operate in the single-loop manner.

One must conclude that the best team practices found in industry today are "double-loop" types without the term ever being mentioned!

9.5 Developing into a Learning Organization

This chapter began with the statement that Learning Organizations could very well be the last frontier for business survival. Is this true? There is absolutely no evidence to support this view, but one may argue that the principles defined for Learning Organizations are fairly new and it may take another two decades before an analysis could make any reliable conclusion. Meanwhile, those organizations purporting to be Learning Organizations should begin to display the successful characteristics. For example, those organizations working with SMTs have not been slow in trumpeting their successes – but what we need is data to show the superior economic performances and growth compared to other organizations operating in the same industry. One would expect some increases in manpower (but lagging economic growth) with minimal turnover compared to the parallel industrial norm.

But so far, none of this is proven nor demonstrated to occur. The main problem is that current team activities are also generating incredible results, so why should one move towards Learning Organizations in the first place?

Maybe the arguments that Kaufman, Senge and Handy had made, refer to 'those other organizations' and not to us – so the thinking goes – and such companies are probably correct in their conclusion.

In a recently published review article by Rousseau (1997), Learning Organizations are casually presented as just *one* of the many areas of interest in "Organizational Behavior in the New Organizational Era". However, the arguments presented for its development are very persuasive, and seem to be superior to the more recent successful managerial approaches such as TQM and Reengineering.

It may be sufficient for us to argue that the Learning Organization approach can lead to superior performances and is, therefore, worth developing in its own right. Leave aside the claim that it is the only approach that can best survive into the future; how then, could we achieve the status of a true Learning Organization?

9.5.1 Developing teams for the Learning Organization. I begin by assuming that Peter Senge, has written the ultimate word on Learning Organizations, with nothing more to add. This is my starting point – our main task then is to review how this ideal state can be achieved.

Much depends on the 'starting point' of a company wanting to become a Learning Organization. Its management style must be *somewhere* when it begins the move. It is most unlikely that an authoritarian-run company will suddenly decide to become a Learning Organization, since its current leadership style would be totally alien to the move. But it is known that even such companies may alter their management style with a new ownership or, with a new CEO – but not to a Learning Organization! It would be an organization employing SMTs, or operating with a TQM philosophy, that would likely be the early candidate for a move to a Learning Organization. There are some similarities between the two concepts that would make it easier for the CEO to be motivated to adopt the Learning Organization philosophy.

The 'motivation to adopt' would really be an act of faith for most people. What Peter Senge describes as a Learning Organization (in its entirety) must be **experienced personally** for it to cause a CEO to want to adopt its philosophy; otherwise the CEO merely reads words which can only be understood at the intellectual level. It is something like wanting to enter, what the Hindus describe as the transcendental state of super consciousness, called "samahdi" or "nirvana" – it would be just an 'intellectual exercise' until the real thing is experienced. And once experienced, one will never want to live outside of this state.

Companies starting off from the Reengineering mode of operation, may feel it quite alien to consider adopting the Learning Organization philosophy. Such companies have its champions and heroes, who take credit for all the good things (if any) that Reengineering might bring. The reader must avoid assuming that operating as a Learning Organization would guarantee the company's performance. Whatever current style a company is working with is OK if the leadership prefers it – and it probably makes money as well. Learning Organizations are just going to be a different way of working – with a promise for changes that will supposedly alter their entire attitude to work and life – like operating continuously in Maslow's state of self-actualization. The very concept of this possibility is highly motivating for those many people who are endlessly searching to find meaning to their (working) lives.

It is a given that the CEO and the top management team (TMT) are imbibed in and are adherents of the new philosophy; else no movement towards the learning organization will occur. The only meaningful step is for the CEO and members of the TMT to make the commitment to take this journey. The structure and management's influence of the teams will depend on the CEOs and the TMT's commitment to become a Learning Organization.

However, from an operational point, Teams are the logical place to build the Learning Organization. "Team Learning is vital because Teams, not individuals, are the fundamental learning units in modern organizations. Unless teams can learn, the organization cannot learn" (Senge, 1990).

So we should look into the creation and development of teams, and in this, we have much experience – especially with Self-Managed Work Teams (SMTs) and teams used with reengineering applications.

The model proposed is illustrated by a continuous line joining two points A and B. Point A is somewhat equivalent to what may be found in a Theory X organization, but conditions at point B may be a long way from the characteristics of a Theory Y organization (MacGregor, 1960). Instead, point B represents teams operating according to the 'rules' of Learning Organizations. Positions along line AB represent the status of current team operations for a specific company. In practice, we are likely to find a large variety of team configurations that lay between A and B. Teams that limit themselves to simple improvements such as Methods Engineering, would cluster around point A, whereas SMTs are likely to be higher up the scale towards point B.

What is needed is a mathematical equation that determines where a particular company's team operation lies on the scale. The development of such an equation is fortunately outside the scope of this book, and will be left to future behavioral scientists to work on its development and validation.

An organization, positioned somewhere along line AB, may have an interest in adopting Learning Organization concepts. As education and training proceeds, the teams should begin to reposition themselves towards point B. At other times, however, certain factors (such as, a new CEO, a poor financial report, etc.) may cause a reverse in direction along the scale (as with Florida Power and Lighting), but this is *not* to say that it would be taking a retrograde step. We must always keep in mind that the ultimate reason for a business to exist, is to be profitable in the widest sense – its mission is only a *means* to achieve that goal.

Many paths exist for organizations to eventually achieve the status of a Learning Organization. These depend on numerous factors, such as, the nature of the business, leadership style, size of organization, motivation, and so on. Consideration of these issues go beyond the scope of this book; but what *will* be considered, is how the teams could be structured so as to move towards the direction of point B in Figure 9.2.

9.5.2   Structure of the teams. It is proposed to utilize a team structure similar to that used in Reengineering. But Reengineering itself is NOT on our minds. Under Reengineering, the entire operational structure of an organization's activities is defined as a broad set of interconnected processes. By processes, we mean the process structure of the organization (see Hammer and Champy, 1993). Incoming jobs need to be processed and we pass these through departments in order to do the processing. Reengineering concentrates on the processing activities to the almost exclusion of the departments involved. Depending on the company size, an entire organization's activities, whether in service or manufacturing, can be described by about 6 to 8 major process groups. Teams are assigned to each process group, with the intention of making major improvements in both quality and productivity.

For a manufacturing company, examples of process groups contain all the processes involved between a "starting" and "finishing" point within the process boundaries. Examples of such process groups are:

Group 1. Start: Request for quotation   End: A signed contract
from a potential customer.   with delivery dates (DD).

Group 2. Start: A signed contract.   End: Completed Drawings
                                                and   BOM   (the   Bill   of
                                                Materials).

Group 3. <u>Start</u>: Completed Drawings   <u>End</u>: Complete set of
With the BOM & DD (due date).   needed raw materials in the
                                           warehouse.

Group 4. <u>Start</u>: Delivery dates,   <u>End</u>: Products delivered to
Drawings & BOM, Inventory status.   Customers by DD.

And so on ------

Each process boundary should be wide enough to enable the appropriate team to operate within a broad range of activities. For example, process group '1' involves marketing, R&D, engineering, finance, customer service, plant engineering, logistics and manufacturing; process group '2' involves R&D, engineering, purchasing, logistics, warehousing and transportation; and so on.

These natural process teams should comprise about six persons – half from within the process boundary and half from outside. These process teams do not preclude the creation of other specialized teams, such as, concurrent engineering, QFD, VA/VE, equipment selection, etc. of a technical nature.

It is again emphasized that a Reengineering approach is NOT being proposed, only the <u>structure</u> of its Teams, since between them, the entire organization's activities are covered. This ensures, that at a minimum, there will be a Team to turn to for guidance in resolving any problem that may crop up, in any part of the organization.

Teams are not constrained to only work within their respective boundaries – however, their respective process group would be their minimal area of responsibility where they are expected to search out breakthroughs in quality and productivity. They may use any set of tools they have in their respective 'kit bags' and are encouraged to experiment with potentially good ideas. Teams can actively pursue new concepts in any particular area and are encouraged to interact with any other team in the organization.

Systems thinking and shared vision of the Learning Organization will still guide their mode of operation. The number and scope of the teams may vary with respect to time according to the organizational needs.

Unlike reengineering, they would never countenance the dismissals of 'surplus' workers (especially middle managers) as a part of the solution. Far better for the surplus middle managers to be put to work on developing projects the company has in reserve for future development (see Dar-El,

1997; and Dar-El and Kirshenbaum, 1999). Guiding the process teams would be a Process Steering Team comprised of senior managers and at least one representative of the CEO, whose task would be to assist in absorbing and implementing the many changes that are bound to be made. Above the steering team would be the CEO and the TMT providing the strategic design methodology for developing and sustaining the Learning Organization.

Even with high autonomy, shared vision, systems thinking, and top management commitment, the sum total of these efforts may still not fully classify the company as a Learning Organization. Still needed is a plan to capture the created knowledge – on how to store, retrieve, and generally share this knowledge among members of the organization.

One approach could be to formally identify experienced key personnel who become the 'keepers of knowledge' – to be consulted and tapped at will. Another approach could be to maintain detailed records of all decisions, conditions and the associated environmental situation at the time of the 'happening', which can be tapped during future times. The only problem with the 'written' word is that the factors influencing the behavior of some matter or other, can get to be very complex. In fact, too complex or subtle to do justice via the written word (think of the times a leader makes decisions (especially <u>correct</u> ones) on the basis of a haunch).

It could be more effective (and probably less expensive) to maintain this knowledge in the hands of the company's 'wise old men' who are able to size up the current situation and place it in the context of similar situations of the past. All this does is to provide input for the current decisions to be made. On the other hand, when ideas are created during brainstorming , or other creative problem-solving exercises, there may be several ideas generated, although not adopted at the time, that are potentially excellent and worth recording for future consideration when the same or similar topics are being reviewed. Such ideas need to be recorded for future reference – without in anyway, diminishing the services of the 'wise men'.

Finally, there will be a need to focus on the measurement process. Overall company performance is regularly measured anyway – so this will indicate how the overall concepts are working for the good of the organization. But when we move to a Learning Organization, we must expect much more. What are the factors that lead to a successful implementation? How should these be measured? In what way could we improve the implementation process? How should this be measured? The measurements should compliment the expected time-phased objectives for becoming a Learning Organization.

One cannot overstress the role played by the organization's leader. If the dream is there, then good things can happen – but we're only at the beginning of the journey. The philosophy of the new era must be both propounded and practiced – everyone needs to believe that you mean it! This way, one triggers the imagination and vision for others to follow. Train, Train and Train yet more. Individuals must get into the habit of working in a different way – and practice makes perfect. Be persistent, and as with individual learning, one will become more and more proficient at working in this new and fascinating manner. Once individuals are constantly operating in what may be described as Maslow's, (1970) level of 'self actualization', they may never want to return to the old way of working.

This writer is heavily indebted to Peter Senge for his deep thoughts and exposition given in the "Fifth Discipline". One can only wish that many more would read and be as influenced by this work as I was.

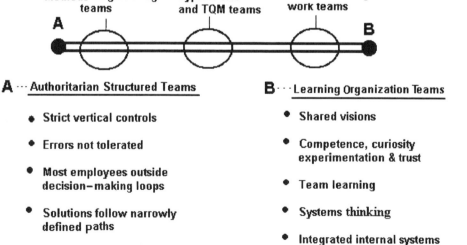

Figure 9.2: A 'scale model' for learning organization teams

10 A SUMMARY AND FUTURE RESEARCH ON HUMAN LEARNING

This book is presented in ten chapters. The bulk of these deal with individual learning. The learning models covered were essentially those that needed the estimation of only two learning parameters, and models requiring three or more (except for Towill) were relegated to "others". Have I done a disservice to the "others"? I think not.

Learning Models are, by definition, to be used as predictors of the future. Learning curves are of no value if we need to pretest the system in order to obtain the data to be analyzed – since this very data is supposed to help design the work system which was already decided upon (in order to get the data)!

In this area, DeJong's model is resuscitated and new life breathed into it. New research needs to confirm that the DeJong model is alive and well, since it has the potential of becoming an exceedingly popular and sturdy learning model.

Parameter estimation is a new contribution to the work on Human Learning. It's the only work done to date that tries to objectively generate the learning parameters. Once we are able to do this, the mystery is removed from learning curves. The relationship developed between ϕ (or 'b') and $\left(\dfrac{C}{t_1} \right)$ is exceedingly important, since it provides designers with another empirically-based strong relationship to use in their learning calculations.

The whole area of learning, forgetting and relearning is new. Not much material is available, but the maximum has been squeezed out from whatever is available to give a reasonable analysis of the situation.

The Cost Model for Optimal Training Schedules is another innovation and is the first time this material appears in a book on Human Learning.

Unfortunately, we did not cover any group/crew learning – there were simply no research papers to cover this type of work. The same is virtually true for Plant Learning and Product Learning. Most of the published articles are of an anecdotal nature, based on 'action research' or, simply uncontrolled industrial experimental work of little value to the serious researcher.

Finally, the introduction into Learning Organizations completes the circle. Learning Organizations may very well become a way of life for our business in the future. We need to know the impact that Human Learning has in this area and what can be done to enhance its development and application. It is still too new to say, but many points pertaining to Learning Organizations are exposed and discussed.

My greatest criticism of this book is the lack of adequate integration between the work of the psychologists and the engineers. Both groups have made important contributions, but I've had the tendency to separate their work and to concentrate mainly on work by the engineer/ergonomist. A future book should integrate the two worlds and blend the approaches into one.

Knowledge is forever expanding. I believe this book is a great advance on any previously published work on Human Learning. I am sufficient a pragmatist to know that the next book that will appear on this subject will take knowledge of the new topics to even greater depths. It is with this view that I now make my comments for future research in Human Learning.

In reviewing this book, it becomes fairly evident that material is based on reliable evidence and those areas that are weak and require further investigation.

The following points summarize the comments I wish to make; these are further classified under the following subtitles:

> 10.1   Individual Learning
> > 10.1.1   Experience
> > 10.1.2   The Learning-Forgetting-Relearning (L-F-R) Phenomenon
> > 10.1.3   Estimating Learning Parameters
> > 10.1.4   Learning with Long Cycle Times/Large Products
> > 10.1.5   Optimal Training Schedules
>
> 10.2   Team/Crew/Group Learning
> 10.3   Plant Learning
> 10.4   The Learning Organization.

10.1 Individual Learning

The learning characteristics of naive, individual operators, without Forgetting, are fairly well understood.   But in marked contrast, and 'shouting' to be heard, is how to assess "experience"!

10.1.1 Experience. Experience can refer to 'similar type' work, work in the general area, or else, identical work. How should experience be assessed? Surely, this aspect of work planning is far more prevalent in industry than the frequency of work planned by naive operators.

Much research is needed for modeling "experience" into the learning curve and to verify the predictions produced from such models.

10.1.2 The Learning-Forgetting-Relearning phenomenon. The next topic in individual learning is the Learning-Forgetting-Relearning (L-F-R) phenomenon. We need to better understand the L-F-R phenomenon, especially applied to cycling processes. In totality, there are only some 3-4 articles written so far on this subject, but it is a topic that could easily involve the work of a hundred research teams. However, the literature is empty – awaiting the expected onslaught!

In terms of relevancy, the L-F-R phenomenon is highly prevalent in real life and whatever model(s) is (are) built can be checked against real data. Chapter 5 of this book has plenty of examples that are relevant to the L-F-R phenomenon.

10.1.3 Estimating learning parameters. We could do with some verification studies, the methods used for evaluating the learning parameters, t_1 and b. The strong relationship between ϕ (or 'b') and $\left(\frac{c}{t_1}\right)$ is such, that one has essentially received a new, easy to evaluate, 'equality' for use in all calculations dealing with Learning Theory. Therefore, it is worth sharpening this relationship.

In particular, we should be most interested in knowing and verifying the values of the learning parameters under conditions of learning-forgetting-relearning.

The DeJong model comes alive and researchers should demonstrate its universal application and ensure that the model is easily used with reliability.

10.1.4 Learning with long cycle times/large products. There is really only one paper dealing with long cycle times. The topic aches for a reviewer and an improved approach for tackling this topic. Long cycle time work is a major part of today's learning experience and is by far the more important issue found in practice. Israeli R&D companies tend to fit this requirement, but who should their planners speak to?

The situation with whole products may even be in a worse situation. Where is a proven model for developing cost models for new products?

Industry relies on a bunch of heuristics – but there are no models available for prediction.

Future researchers should go into the field and find out the important issues from the people that need the information. For example, be sure that major aircraft companies are certain to consider learning aspects for assessing costs on the next aircraft model, since they have a long history for applying learning theory methods. However, the same should apply to many other industries as well

10.1.5 Optimal training schedules. Retraining implies reversing the effects of the forgetting process so that the retention of a task is increased. Chapter 7 is the first attempt (based on my knowledge) at putting this topic in perspective. Much more work is needed by our mathematically-oriented students in searching for simpler ways to tackle this problem.

Training in industry is essential – especially for sensitive operations such as control-based operations. How much training should be invested, and at what frequency? This is a new area for research.

10.2 Team/Crew/Group Learning

This area is entirely missing from this book, simply because our regular researchers have not done any work on the topic. But, it is a rich area for research, both mathematically, as well as from a behavioral point of view. Nevertheless, since so much of our work is done as a group (crew) effort, we should understand and model the learning behavior of such working crews.

10.3 Plant Learning

Thus far, we have accumulated a number of techniques for looking at specific aspects of this problem. But, *no* model for plant learning exists. We tend to be reactive to the firm's past data – then, model the future on the assumption that what was done in the past will continue into the future.

Today, we're just beginning to notice that the application of several improvement processes can have a radical influence on a firm's performance – just how much, we're not so sure, but an influence that cannot be denied. How can we isolate the effect of such inputs into the system? How do we measure the expectations?

People using the PDCA cycle have come out with proven results. Applied to the electronic industry, learning slope parameters in the order of 60% have been achieved! No industrial learning slope comes close to this.

It is without a doubt that the area of process improvement has important ramifications for influencing the learning curves of plants – but we await the models that could help us plan and design such systems. We're already entering the realm of socio-technical design and it is not so easy (but very rewarding) to get into this type of research. Yet, without this, our application is just going to be based on one more 'action research' based on a bunch of heuristic rules that can never be verified.

Finally, some researchers may find it rewarding to try to find some relationship between t_1, b (or ϕ) and the MTM value of a product, much in the same way as was done for Individual Learning. We need to develop some capability for predictability to plan with progress functions.

10.4 The Learning Organization

This is a more problematic area for this writer.

The first step is a measurement scale that can approximately determine 'how close' an organization is to a Learning Organization. This is no problem for the fainthearted to tackle – it needs to be done, but which behavioral group will develop the methodology and measuring scale?

Teams form the basic elements of a Learning Organization. How should we measure the effectiveness of such teams? Earlier, we wrote of 'team/crew/group' learning and this fits very nicely with our work experience, since many industrial tasks need to be accomplished in teams (e.g., making tires, working in an assembly line, maintenance teams for oil refineries, and so on). Now we are referring to teams that are directed towards achieving some general improvement goals., e.g., Quality Circles, SMT, WITs, and so on. But Learning Organizations make a distinction between these teams and those operating under the philosophy of the Learning Organization. How can we distinguish between these two? Also, who will develop the measures?

Outwardly, one gets the impression that Human Learning 'stops' at the individual learning and/or plant learning level, and that a new set of learning is needed to make sense out of Learning Organizations. However, it still is the same individuals who comprise the teams in a Learning Organization environment and it is upon these persons that we need to focus.

Learning Organizations is a relatively new concept and it will still take some time for these concepts to gel into a meaningful methodology. With the passing of time, I am convinced that much will be written and developed on this subject – then, it will become clearer in which direction and emphasis research should take.

REFERENCES

Abell, D.F. and Hammond, J.S. (1979), *Strategic Market Planning*, Prentice-Hall, Englewood Cliffs, NJ, USA.

Abernathy, W.J. and Wayne, K. (!974), "Limits of the learning curve", *Harvard Business Review*, 52(5), 109-119.

Abernathy, W.J., Clark, K.B. and Kantorow, A.M. (1981), "The new industrial competition", *Harvard Business Review*, 59, 68-81.

Adams, J.A. (1987), "Historical review and appraisal of research on the learning, retention and transfer of human motor skills", *Psychological Bulletin*, 101(1), 41-74.

Adler, G.L. and Nanda, R. (1974a), "The effects of learning on optimal lot size determination - the single product case", *AIIE Trans.*, 6(1), 14-20.

Adler, G. and Nanda, R. (1974b), "The effects of learning on optimal lot size determination - multiple product case", *AIIE Trans.*, 6(1), 21-27.

Adler, P.S. (1990), "Shared learning", *Management Science*, 36(8), 938-957.

Adler, P.S. and Clark, K.B. (1991), "Behind the learning curve: a sketch of the learning process:, *Management Science*, 37(3), 267-281.

Akao, Y. (1990), *Quality Function Deployment: Integrating Customers Requirements into Product Design*, Productivity Press, Portland, OR, USA.

Alchain, A. (1950), "Reliability of progress curves in airframe production", *Econometrica*, 31, 679-693.

Alden, J.R. (1974), "Learning curve and example", *Industrial Engineering*, 6(12), p. 34.

Allemang, R.D. (1977), "New technique could replace learning curves", *Industrial Engineering*, 9(8), 22-25.

Anderlohr, G. (1969), "What production breaks cost", *Industrial Engineering*, 1, 34-36.

Andress, F. (1954), "The learning curve as a prediction tool", *Harvard Business Review*, 32, 87-97.

Anzai, Y. and Simon, H.A. (1979), "The theory of learning by doing", *Psychological Review*, 86, 124-140.

Argote, L., Beckman, S. and Epple, D. (1990), "The persistence and transfer of learning in industrial settings", *Management Science*, 36(2), 140-154.

Argyris, C. (1977), "Double loop learning in organizations", *Harvard Business Review*, 55(5), 115-125.

Argyris, C. (1991), "Teaching smart people how to learn", *Harvard Business Review*, 69, 99-109.

Argyris, C. (1992), *On Organizational Learning*, Blackwell Pub., Cambridge, MA, USA.

Argyris, C. and Schoen, D.A. (1996), *Organizational Learning II*, Addison-Wesley, Reading, MA, USA.

Arrow, K.J. (1962), "The economic implications of learning by doing", *Review of Economic Studies*, 29, 166-170.

Arzi, Y. and Shtub, A. (1997), "Learning and forgetting in mental and mechanical tasks: a comparative study", *IIE Trans.*, 29, 759-768.

Ashton, J.E., Fagan, R.L. and Cook, F.X. (1990), "From status quo to continuous improvement: the management process", *Manufacturing Review*, 3(2), 85-90.

Asseo, R. (1987), Learning and Forgetting Curve Analysis and Combination. Unpublished MSc Thesis, Tel Aviv University, Israel (in Hebrew).

Ayas, K. and Foppen, J.W. (eds.) (1996), *The Learning Organization and Organizational Learning*, The Rotterdam School of Management, The Netherlands.

Badiru, A. (1992), "Computational survey of univariate and multivariate learning curve models", *IEEE Trans. on Engineering Management*, 39(2), 176-188.

Badiru, A.B. (1994), "Multifactor learning and forgetting models for productivity and performance analysis", *Int. J. Human Factors in Manufacturing*, 4(1), 37-54.

Badiru, A.B. (1995), "Multivariate analysis of the effect of learning and forgetting on product quality", *Int. J. Production Research*, 33, 777-794.

Bailey, C.D. (1989), "Forgetting and the learning curve: a laboratory study", *Management Science*, 35(3), 346-352.

Baloff, N. (1966), "Startups in machine-intensive production systems", *J. of Industrial Engineering*, 17, 25-32.

Baloff, N. (1966 b), "The learning curve - some controversial issues", *The J. Industrial Economics*, 15, 275-282.

Baloff, N. (1970), "Startup management", *IEEE Trans. of Engineering Management*, EM-17, 132-141.

Baloff, N. (1971), "Extensions of the learning curve - some empirical results", *Operational Research Quarterly*, 22(4), 329-305.

Bapu, H. (1967), Learning Curves for Punch Press Operators. Project Report to S. Kooz, May, 1967.

Barany, J.A. (1982), "Reconsider learning allowances when setting time standards". Unpublished research paper communicated to Dar-El.

Barnes, R. and Amrine, H. (1942), "The effect of practice on various elements used in screw-driver work", *J. of Applied Psychology*, 26, 197-209.

Barnes, R. (1980), *Motion and Time Study* (8th edition), Wiley and Sons, New York, USA.

Barrow, J.A.W. (1993), "Does total quality management equal organizational learning?", *Quality Progress*, 26, 39-43.

Bass, F.M. (1980), "The relationship between diffusion rates, experience curves, and demand elasticies for consumer durable technical innovations", *J. of Business*, 53(3), 551-567.

Baybars, I. (1986), "Survey of exact algorithms for simple assembly line balancing", *Management Science*, 32(8), 909-932.

Behnezhad, A.R. and Khoshnevis, B. (1988), "The effects of manufacturing progress function on machine requirements and aggregate planning problems", *Int. J. Production Research*, 26(2), 309-326.

Belhaoui, A. (1986), *The Learning Curve*, Quorum Books, Westport, CT, USA.

Berndt, E. (1991), *The Practice of Econometrics - Classic and Contemporary*, Addison-Wesley Publishing Co., USA.

Bevis, F.W., Finnear, C. and Towill, D.B. (1970), "Prediction of operator performance during learning of repetitive tasks", *Int. J. Production Research*, 8(4), 293-305.

Billon, S.A. (1966), "Industrial learning curves and forecasting", *Management International Review*, 6(6), 65-96.

Bohlen, G.A, and Baraney, J.W. (1976), "A learning curve prediction model for operators performing industrial bench assembly operations", *Int. J. Production Research*, 14(2), 295-303.

Boston Consulting Group, (1970), Perspectives on Experience. Boston Consulting Group, Boston, MA, USA.

Brenner, M. (1990), Prediction of Learning in Short Cycle Tasks. Unpublished MSc Thesis, Technion, Israel (in Hebrew).

Bryan, W.L. and Haster, N. (1899), "Studies on the telegraphic language, the acquisition of a hierarchy of habits", *Psychological Review*, 6, 345-375.

Buck, J.R., Tanchoco, J.M.A. and Sweet, A.L. (1976), "Parameter estimation methods for discrete exponential learning curves", *AIIE Trans.*, 8(2), 184-194.

Buck, J.R. and Cheng, S.W.J. (1993), "Instructions with feedback effects on speed and accuracy with different learning curve models", *IIE Trans.*, 25(6), 34-47.

Camm, J.D. (1985), "A note on learning parameters", *Decision Sciences*, 16, 325-327.

Camm, J.D., Evans, J.R. and Womer, N.K. (1987), "The unit learning curve approximation of total cost", *Computers and Industrial Engineering*, 12(3), 205-213.

Carlson, J.G.H. and Rowe, A.J. (1976), "How much does forgetting cost?", *Industrial Engineering*, 8(9), 40-47.

Carlson, J.G.H. (1962), "Production standards for small-lot manufacturing", *J. of Industrial Engineering*, 13(6), 496-502.

Carr, G.W. (1946), "Peacetime cost estimating requires new learning curves", *Industrial Aviation*, XX, 76-77.

Carroll, B. (1998), "The self-management payoff: making ten years of improvement in one", *National Productivity Review*, 18(1), 21-27.

Carron, A.V. (1971), *Laboratory Experiments in Motor Learning*, Prentice-Hall Inc., Englewood Cliffs, NJ, USA.

Caspari, J. (1972), Learning Curves, The Israel Institute of Productivity, Tel Aviv (in Hebrew).

Chaffin, D.B. and Hancock, W.M. (1967), Factors in Manual Skill Training, MTM Association for Standards and Research, Report #114.

Chakravarty, A.K. and Shtub, A. (1985), "Balancing mixed-model lines with in-process inventories", *Management Science*, 31, 1161-1174.

Chakravarty, A.K. and Shtub, A. (1986), "A cost minimization procedure for mixed model production lines with normally distributed task times", *European J. of Operational Research*, 23, 25-36.

Chakravarty, A.K. and Shtub, A. (1986), "Dynamic manning of long cycle assembly lines with learning effects", *IIE Trans.*, 18, 392-397.

Chakravarty, A.K. and Shtub, A. (1992), "The effect of learning on the operation of mixed-model assembly lines", *Production and Operations Management*, 1(2), 198-211.

Chassen, J. (1945), "Estimating direct labor cost in aircraft production", *Industrial Aviation*, XX, 56-63.

Chawla, S. and Renesch, J. (eds.) (1995), *Learning Organizations*, Productivity Press, Portland, OR, USA.

Chen, J.T. (1983), "Modeling learning curve and learning complementarily for resource allocation and production scheduling", *Decision Sciences*, 14, 170-186.

Cherrington, J.E. and Towill, D.R. (1980), "Learning performance of industrial long cycle time group task", *Int. J. Production Research*, 18(4), 411-425.

Chiu, H.N. (1997), "Discrete time-varying demand lot-sizing models with learning and forgetting effects", *Production Planning and Control*, 8(5), 484-493.

Christiaansen, R.E. (1980), "Prose memory: forgetting rates for memory codes", *J. Experimental Psychology: Human Learning and Memory*, 6(9), 611-619.

Clarke, F.H., Darrough, M.N. and Heineke, J.M. (1982), "Optimal pricing policy in the presence of experience effects", *J. of Business*, 55(4), 517-530.

Cochran, E.B. (1960), "New concepts of the learning curve", *J. of Industrial Engineering*, 11(4), 317-327.

Cochran, E.B. (1968), *Planning Production Costs: Using the Improvement Curve*, Chandler Pub. Co., San Francisco, CA, USA.

Cochran, E.B. (1969), "Learning: new dimension in labor standards", *Industrial Engineering*, 1(1), 38-47.

Cochran, E.B. and Sherman, H.A. (1982), "Predicting new product labor hours", *Int. J. Production Research*, 20(4), 517-543.

Cohen, L. (1995), *Quality Function Deployment: How to Make QFD Work for You*, Addison-Wesley, Reading, MA, USA.

Cohen, Y. (1992), Optimization of the Makespan in Production Lines Under Learning. Unpublished MSc Thesis, Technion, Israel (in Hebrew).

Cohen, Y. and Dar-El, E.M. (1998), "Optimizing the number of stations in assembly lines under learning for limited production", *Production Planning and Control*, 9(3), 230-240.

Conley, P. (1976), "Experience Curves as a planning tool", in *Corporate Strategy and Product Innovation*, Rothberg, R. (ed.), Free Press, New York, USA.

Conway, R. and Schultz, A. (1959), "The manufacturing process function", *J. of Industrial Engineering*, 10(1), 39-54.

Cooke, N.J., Durso, F.T. and Schvaneveldt, R.W. (1994), "Retention of skilled search after nine years", *Human Factors*, 36, 597-605.

Crawford, J.R. (1944), "Statistical accounting procedures in aircraft production", *Aeronautical Digest*, 44, p. 78.

Crawford, J.R. and Strauss, E. (1947), Crawford-Strauss Study. Air Material Command, Dayton, OH, USA.

Crossman, E.R.F.W. (1959), "A theory of the acquisition of speed-skills", *Ergonomics*, 2(2), 153-166.

Cunningham, J.A. (1980), "Using the learning curve as a management tool", *IEEE Spectrum*, 17(6), 45-48.

Dada, M. and Srikanth, K.N. (1990), "Monopolistic pricing and the learning curve: an algorithmic approach", *Operations Research*, 38(4), 656-666.

Dance, D. and Jarvis, R. (1992), "Using yield models to accelerate learning curve progress", *IEEE Trans. on Semiconductor Manufacturing*, 5(1), 41-46.

Dar-El, E.M. (1960) Consulting Experience with the Housing Building Commission, Melbourne, Victoria, Australia.

Dar-El, E.M. and Rubinovitz, J. (1979), "MUST - a multiple solutions technique for balancing single model assembly lines", *Management Science*, 25, 1105-1114.

Dar-El, E.M. (1993), "EFT-earned free time", *Proceedings of the Productivity and Quality Improvement in Government*. John S.W. Farger (ed.), pp. 166-170, Industrial Engineering and Management Press, Norcross, GA, USA.

Dar-El, E.M. (1986), *Productivity Improvement: Employee Involvement and Gainsharing Plans*, Elsevier, The Netherlands.

Dar-El, E.M. and Rabinovitch, M. (1988), "Optimal planning and scheduling of assembly lines", *Int. J. Production Research*, 26(9), 1433-1450.

Dar-El, E.M. (1990) - Consulting contacts with Ashot Ashqelon, Ashqelon, Israel.

Dar-El, E.M. (1991), Design of Assembly Systems (DAS), Workshop Paper #1, 11th Int. C. Production Research (ICPR), Hefei, China.

Dar-El, E.M. and Rubinovitz, J. (1991), "Using learning theory in assembly lines for new products", *Int. J. of Production Economics*, 25, 103-109.

Dar-El, E.M. (1992), Consulting Experience with a Company in the Food Industry.

Dar-El, E.M. (1993), Consulting Experience with Kibbutz and Koor (trade union) Industries.

Dar-El, E.M. and Altman, H. (1994), "A dynamic-b model of learning", *Proceedings of the Fourth International FAIM (Flexible Automation and Integrated Manufacturing) Conference*, pp. 749-758, Blacksburg, VA, USA.

Dar-El, E.M. (1995), Lectures on TPQM (Total Productivity and Quality Management).

Dar-El, E.M., Ayas, K. and Gilad, I. (1995a), "A dual-phase model for the individual learning process in industrial tasks" *IIE Trans.*, 27(3), 265-171.

Dar-El, E.M., Ayas, K. and Gilad, I. (1995b), "Predicting Performance times for long cycle time tasks", *IIE Trans.*, 27(3), 272-281.

Dar-El, E.M. and Zohar, D. (1998), Training for Emergency: Retention of Skills for Acting During an Emergency. Government Safety Committee, Jerusalem, Israel.

Dar-El, E.M. and Kirshenbaum, A. (1999), " 'Theory M': A New Strategy for Quality and Productivity Improvement". Under review.

Dasgupta, P. and Stiglitz, J. (1988), "Learning by doing market structure, industrial and trade policies", *Oxford Economic Papers*, 40, 246-268.

Day, G.S. and Montgomery, D.B. (1983), "Diagnosing the experience curve", *J. Marketing*, 47, 44-58.

DeJong, J.R. (1957), "The effects of increasing skill on cycle time and its consequences for time standards", *Ergonomics*, 1(1), 51-60.

DeJong J.B. (1964), "Increasing skill and reduction of work time", *Time and Motion Study*, 13(9), 28-41.

Deming, W.E. (1982), Quality, Productivity and Competitive Position. M.I.T. Center for Advanced Engineering Study, Cambridge, MA, USA.

Deming, W.E. (1986), Out of the Crisis. M.I.T., Center of Advanced Engineering Study, Cambridge, MA, USA.

Dolan, R.J. and Jeuland, A.P. (1981), "Experience curves and dynamic demand models: implications for optimal pricing strategies", *J. of Marketing*, 45(1), 52-62.

Donath, N., Globerson, S. and Zang, I. (1981), "A learning curve model for a multiple batch production process", *Int. J. Production Research*, 19(2), 165-175.

Donovan, J.M. (1986), "Self-managed work teams - extending the quality circle concept", *Quality Circles Journal*, 9(3), 15-20.

Dorroh, J.R., Gulledge, T.R. and Womer, N.K. (1986), "A generalization of the learning curve", *European J. of Operations Research*, 26(2), 205-216.

Duarte, D. and Snyder, N. (1997), "From experience: facilitating global organizational learning in product development at Whirlpool Corporation", *J. Product Innovation Management*, 14, 48-55.

Dudley, N.A. (1968), *Work Measurement: Some Research Study*, MacMillan, London, UK.

Duncan, J. and Weiss, A. (1976), "Organizational learning: implications for organizational design", pp. 75-123, in *Research in Organizational Behavior*, Stew, B. (ed.), JAI Press, Greenwich, CT, USA.

Dutton, J.M. and Thomas, A. (1984), "Treating progress functions as a managerial opportunity", *Academy of Management Review*, 9(2), 235-247.

Dutton, J.M., Thomas, A. and Butler, J.E. (1984b), "The history of progress functions as a managerial technology", *Business History Review*, 58(2), 204-233.

Ebbinghaus, H. (1913), *Memory*, H. Ruer and C. Bussenius, New York: NY.

Ebert, R.J. (1972), "Time horizon: implications for aggregate scheduling effectiveness", *AIIE Trans.*, 4(4), p. 298.

Ebert, R.J. (1984), "Aggregate planning with learning curve productivity", *Management Science*, 23(2), 171-182.

Eckert, R.L. (1984), "Codes and classification systems", in *Group Technology*, Hyer, N. (ed.), Society of Manufacturing Engineers, Dearborn, MI, USA.

Eggemeier, F.T. and Fisk, A.D. (1992), Automatic Information Processing and High Performance Skills. PC A08/MF A02, Dayton University Ohio Research Institute, USA.

Ekstrand, B.R. (1967), "Effect of sleep on memory", *J. of Experimental Psychology*, 75, 64-72.

Ellis, S. and Globerson, S. (1996), "Diagnosing learning in project management", *Int. J. of Industrial Engineering*, 3(2), 86-94.

Elmaghraby, S.E. (1990), "Economic manufacturing quantities under conditions of learning and forgetting", *Production Planning and Control*, 1(4), 196-208.

Elmuti, D. (1996), "Sustaining high performance through self-managed work teams", *Industrial Management*, 38(2), 4-9.

Engwall, R.L. (1992), "The Learning Curve", pp. 5.145-5.161, in *Maynard's Industrial Engineering Handbook* (4th edition), Hodson, W. (ed.), McGraw-Hill, New York, USA.

Enis, B.M. (1980), "GE. PIMS. BCG, and the PLC", *Business*, 30(3), 10-18.

Erdal, E. (1998), 'Accuracy vs. errors' for psychomotor assembly activities. Private communication.

Fein, M. (1973), "Work measurement and wage incentives", *Industrial Engineering*, 5(9), 49-51.

Fein, M. (1981), *IMPROSHARE: An Alternative to Traditional Managing*, Mitchell Fein, Hillsdale, NJ, USA.

Feller, W. (1961), "Specific interpretations of learning by doing", *J. Economic Theory*, 12, 23-27.

Fine, C.H. (1986), "Quality improvement and learning in productive systems", *Management Science*, 32(10), 1301-1315.

Finn, D.W. (1984), "Productivity predictors of assembly workers using standards based on learning curves", *OMEGA*, 12(6), 569-574.

Fiol, C.M. and Lyles, M.A. (1985), "Organizational learning", *Academy of Management Review*, 10(4), 803-813.

Fisher, J. (1974), *Energy Crisis in Perspective*, Wiley-Interscience, New York, NY, USA.

Fisk, A.D., Ackerman, P.L. and Schneider, W. (1987), "Automatic and controlled processing theory and its applications to human factors problems", *Human Factors Psychology*, 5, 159-197.

Fisk, A.D. and Hedge, K.A. (1992), "Retention of trained performance in consistent mapping search after extended delay", *Human Factors*, 34, 147-164.

Fisk, J.C. and Ballou, D.P. (1982), "Production lot sizing under a learning effect", *IIE Trans.*, 14, 257-264.

Fitts, P.M. (1966), "Cognitive aspects of information processing III: set for speed versus accuracy", *J. of Experimental Psychology*, 71, 849-857.

Fitts, P.M. and Posner, M.I. (1967), *Human Performance*, Brook/Cole Pub. Co., CA, USA.

Fleishman, E.A. and Parker, J.F. (1962), "Factors in the retention and relearning of perceptual-motor skills", *J. of Experimental Psychology*, 64(3), 215-220.

Fleishman, E.A. and Rich, S. (1963), Role of kinesthetic and spatial visual abilities in perceptual motor learning", *J. of Experimental Psychology*, 6, 6-11.

Fudenberg, D. and Tirole, J. (1983), "Learning-by-doing and market performance", *Bell J. of Economics*, 15, 522-530.

Garg, A. and Milliman, D. (1961), "The aircraft progress curve modified for design changes", *J. of Industrial Engineering*, 12(1), 23-28.

Garvin, D. (1988), *Managing Quality*, Free Press, New York, USA.

Garvin, D. (1993), "Building a learning organization", *Harvard Business Review*, 72, 78-91.

Gershoni, H. (1971), "The influence of motivation and micro-method on learning manual tasks", *Work Studies and Management Services*, 15(9), 585-594.

Gershoni, H. (1979), "An investigation of behavior changes of subjects learning manual tasks", *Ergonomics*, 22(11), 1195-1206.

Gilad, I. (1998). Planning of experiments supporting the 'cost of training'. Technion Research Project funded by the Safety Committee, Jerusalem.

Ginzberg, S. (1998), Skill Retention and Learning. Unpublished MSc thesis, Technion, Israel (in Hebrew).

Gitlow, H.S. and Shelly, J. (1987), *The Deming Guide to Quality and Competitive Position*, Prentice-Hall, Edgewood Cliffs, NJ, USA.

Giunipero, L.C. and Vogt, J.F. (1997), "Empowering the purchasing function: moving to team decisions", *Int. J. of Purchasing and Materials*, 33, 8-15.

Globerson, S. (1974), Temporal Aspects of Job Analysis: a Quantitative Approach. Unpublished Ph.D. dissertation, University of California, Berkeley.

Globerson, S. and Crossman, E.R.F.W. (1976), "Minimization of worker induction and training costs through job enrichment", *Int. J. Production Research*, 14(3), 345-355.

Globerson, S. (1980), "The influence of job related variables on the predictability power of three learning curve models", *AIIE Trans.*, 12(1), 64-69.

Globerson, S. (1980), "Introducing the repetition pattern of a task into its learning curve", *Int. J. Production Research*, 18(2), 143-152.

Globerson, S. (1984), "The deviation of actual performance around learning curve models", *Int. J. Production Research*, 22(1), 51-62.

Globerson, S. and Shtub, A. (1984), "The impact of learning curves on the design of long cycle time lines", *Industrial Management*, 26(3), 5-10.

Globerson, S. and Levin, N. (1987), "Incorporating forgetting into learning curves", *Int. J. of Operations and Production Management*, 7(4), 80-94.

Globerson, S. and Riggs, J. (1988), "The effects of imposed learning curves on performance improvements", *IIE Trans.*, 20, 317-324.

Globerson, S. and Seidmann, A. (1988), "The effects of imposed learning curves on performance improvements", *IIE Trans.*, 20(3), 317-324.

Globerson, S., Levin, N. and Shtub, A. (1989), "The impact of breaks on forgetting when performing a repetitive task", *IIE Trans.*, 21(4), 376-381.

Globerson, S. and Millen, R. (1989), "Determining learning curves in group technology settings", *Int. J. Production Research*, 27(10), 1653-1664.

Globerson, S. and Levin, N. (1995), "A learning curve model for an equivalent number of units", *IIE Trans.*, 27, 716-721.

Globerson, S. and Ellis, S. (1995), "Organizational learning analysis of a High Tech company", pp. 1-3, in the *Proceedings of the Decision Sciences Institute*, Boston, MA. USA.

Globerson, S. and Gold, D. (1997), "Statistical attributes of the power learning curve model", *Int. J. Production Research*, 35(3), 699-711.

Globerson, S. Nahumi, A. and Ellis, S. (1998), "Rate of forgetting for motor and cognitive tasks", *Int. J. of Cognitive Ergonomics*, 2(3), 181-191.

Glover, J.H. (1965), "Manufacturing progress functions: I - an alternative model and its comparison with existing functions", *Int. J. Production Research*, 4(4), 279-300.

Glover, J.H. (1966), "Manufacturing progress functions: II - Selection of trainees and control of their progress", *Int. J. Production Research*, 5(1), 43-59.

Glover, J.H. (1967), "Manufacturing progress functions III: Production control of new products", *Int. J. Production Research*, 6(1), 15.

Goel, S.N. and Becknell, R.H. (1972), "Learning curves that work", *Industrial Engineering*, 4(5), 28-31.

Goetsch, D.L. and Davis, S.B. (1998), *Understanding and Implementing ISO-9000 and ISO Standards*, Prentice-Hall, Upper Saddle River, NJ, USA.

Gold, B. (1981), "Changing perspectives on size, scale, and returns: an interpretive survey", *J. of Economic Literature*, 19, 5-33.

Goldberg, J.H. and O'Rourke, S.A. (1989), "Prediction of skill retention and retraining from initial training", *Perceptual and Motor Skills*, 69(2), 535-546.

Golea, M. and Marchand, M. (1993), "Learning curves of the clipped Hebb rule for networks with binary weights", *J. Physics A: Math. Gen.*, 26, 5751-5766.

Goonetilleke, R.S., Drury, C.G. and Sharit, J. (1995), "What does an operator need to learn?", pp. 1284-1288, in *Proceedings of the 39th Annual Meeting of the Human Factors and Ergonomics Society*.

Gopher, D. and Donchin, E. (1986), "Workload: an examination of the concept", in *Handbook of Perception and Human Performance*, Boff, R.K., Kaufman, L. and Thomas, J.P. (eds.).

Goralnick, I. (1998), Optimal Assembly Balancing under Learning. Incomplete MSc thesis, Technion, Israel.

Gray, T.H. (1979), Boom operator part task trainer: Test and evaluation of the transfer of training. AFHRL, Williams Air Force Base, AZ, USA.

Gruber, H. (1992), "The learning curve in the production of semiconductor memory chips", *Applied Economics*, 24, 885-894.

Gu, P. and Kusiak, A. (eds.) (1993), *Concurrent Engineering: Methodology and Applications*, Elsevier, Amsterdam, The Netherlands.

Gulledge, T.R., Thomas, A. and Dorroh, J.R. (1984), "Learning and costs in airframe production: a multiple output production function approach", *Naval Research Logistics Quarterly*, 31, 67-85.

Gulledge, T.R. and Khoshnevis, B. (1987), "Production rate, learning, program costs: survey and bibliography", *Engineering Costs and Production Economics*, 11, 223-236.

Gulledge, T.R. and Womer, N.K. (1990), "Learning curves and production functions: an integration", *Engineering Costs and Production Economics*, 20, 3-12.

Gulledge, T.R., Tarimcilar, M.M. and Womer, N.K. (1997), "Estimation problems in rate-augmented learning curves", *IEEE Trans. on Engineering Management*, 44(1), 91-97.

Hackett, E.A. (1983), "Application of a set of learning curve models to repetitive tasks", *Radio and Electronic Engineer*, 53(1), 25-32.

Hagman, D.J. and Rose, A.M. (1983), "Retention of military tasks: a review", *Human Factors*, 27(2), 199-213.

Hall, W.K. (1980), "Survival strategies in a hostile environment", *Harvard Business Review*, 58(5), 75-85.

Hammer, M.A. (1990), "Reengineering work: don't automate, obliterate", Harvard Business Review, 68(4), 104-112.

Hammer M.A. and Champy, J. (1993), *Reengineering the Corporation: A Manifesto for Business Revolution*, Harper Business, New York, NY, USA.

Hammer M.A. (1996), *Beyond Reengineering: How the Process-Centered Organization is Changing*, Harper Business, New York, NY, USA.

Hancock, W.M. and Foulke, J.A. (1963), Learning Curve Research on Short Cycle Operations: Phase I, Laboratory Experiments, MTM Association for Standards and Research, Report #112.

Hancock, W.M., Clifford, R.R., Foulke, J.A. and Krystynak, L.F. (1965), Learning Curve Research on Manual Operations: Phase II, Industrial Studies, MTM Association for Standards and Research, Report #113.

Hancock, W.M. (1967), "The prediction of learning rates for manual operations", *J. of Industrial Engineering*, 18(1), 42-47.

Hancock, W.M. (1971), "The Learning Curve", Ch. 7, 102-114, in *The Industrial Engineering Handbook* (3rd edition), Maynard, H.B (ed.-in-chief), McGraw Hill Book Co., New York, USA.

Hancock, W.M. and Sathe, P. (1989), Learning Curve Research on Manual Operations, The MTM Associations, Research Report #113A, Fairlawn, NJ, USA.

Hancock, W.M. and Bayah, F.H. (1992), "The Learning Curve" pp. 1585-1598 in *The Handbook of Industrial Engineering* (2nd ed.), G. Salvendy (ed.), Wiley, New York, USA.

Hartley, K. (1965), "The learning curve and its application to the aircraft industry", *J. Industrial Economics*, 13(2), 122-128.

Harvey, R.A. and Towill, D.R. (1981), "Application of learning curves and progress functions: past, present and future", in *Industrial Applications of Learning Curves and Progress Functions*, Towill, D.R. and Harvey, R.A.(eds.), Institute of Electronics and Radio Engineers, London, UK.

Harvey, R.A. (1981), "Analysis of contributory factors in aircraft production learning", pp. 52, in *Proceedings of IERE's Int. C. on the Application of Learning Curves and Progress Functions*, London, '81.

Hatchings, B. and Towill, D.R. (1975), "An error analysis of the time constant learning curve model", *Int. J. Production Research*, 13(2), 105-135.

Haworth, R.P. (1996), "Understanding individual learning for organizational learning", pp. 791-795, in the *Proceedings of the Human Factors and Ergonomics Society*, 40th Annual Meeting.

Hayes, R.H. and Wheelwright, S.C. (1979), "The dynamics of process-product life cycles", *Harvard Business Review*, 57(2), 127-136.

Hayes, R.H. (1981), "Why Japanese factories work", *Harvard Business Review*, 79, 56-66.

Hayes, R.H. and Wheelwright, S.C. (1984), *Restoring Our Competitive Edge: Competing Through Manufacturing*, Wiley and Sons, New York, NY, USA.

Hayes, R.H. and Clark, K.B. (1985), "Exploring the source of productivity differences at the factory level", in *The Uneasy Alliance: Managing the Productivity-Technology Dilemma*, Clark, K.B., Hayes, R.H. and Lorenz, C. (eds.), Harvard Business School Press, Boston, MA, USA.

Heally, A., Clowson, D., MacNamara, D., Marmie, W., Schneider, V., Richard. T., Crutcher, R., King, C., Ericsson, K. and Broune, L. (1995), "The long-term retention of knowledge and skills", *The Psychology of Learning and Motivation*, 30, 135-164.

Hedberg, B. (1981), "How do organizations learn and unlearn?", pp. 8-27, in *Handbook of Organizational Design*, Nystrom, P. and Starbuck, H. (eds.), Oxford University Press, London, UK.

Heineke, J.M. (1986), "Notes on estimating experience curves: economic issues", *IEEE Trans. on Engineering Management*, EM-33(2), 113-119.

Hewitt, D., Sprange, K., Yearout, R., Lisnerski, D. and Sparks, C. (1992), "Effects of unequal relearning rates on estimating forgetting parameters associated with performance curves", *Int. J. of Industrial Ergonomics*, 10(3), 217-224.

Hill, F.M. (1996), "Organizational learning for TQM through quality circles", *The TQM Magazine*, 8(6), 53-57.

Hill, S. (1995), "From quality circles to total quality management", pp. 33-53, in *Making Quality Critical*, Wilkinson, A. and Wilmont, H. (eds.), Routledge and Kegan Paul, London, UK.

Hillier, R.S. and Shapiro, J.F. (1986), "Optimal capacity expansion planning when there are learning effects", *Management Science*, 32(9), 1153-1163.

Hirsch, W.Z. (1952), "Manufacturing progress functions", *Review of Economics and Statistics*, 34(2), 143-155.

Hirsch, W.Z. (1956), "Firm progress ratios", *Econometrica*, 24, 136-144.

Hirschman, W.B. (1973), "Profit from the learning curve", *Harvard Business Review*, 52(1), 125-139.

Hitchcock, N.A. (1993), "Can self-managed teams boost your bottom line?", *Modern Materials Handling*, 48(2), 57-59.

Hofer, C.W. and Schendel, D. (1978), *Strategy Formulation: Analytical Concepts*, West, St. Paul, MI, USA.

Hoffman, T.R. (1968), "Effect of prior experience on learning curve parameters", *J. of Industrial Engineering*, 19(8), 412-413.

Hollander, S. (1965), *The Sources of Increased Efficiency: A Study of Dupont Rayon Plants*, M.I.T. Press, Cambridge MA, USA.

Hovland, C.I. (19??), "Human Learning and Retention", Ch. 17 of *Handbook of Experimental Psychology*, Stephens, S.S., Wiley and Sons, USA.

Howell, S.D. (1980), "Learning curves for new products", *Industrial Marketing Management*, 9(2), 97-99.

Howell, W.C. and Kreidler, D.L. (1963), "Information processing under contradictory instructional sets", *J. of Experimental Psychology*, 65, 39-46.

Hoyle, D. (1998), *ISO-9000 Quality Service Handbook* (3rd ed.), Butterworth-Heinmann, Oxford, UK.

Huber, G.P. (1991), "Organizational learning: the contributing processes and the literature", *Organization Science*, 2, 88-115.

Hufbauer, G. (1966), *Synthetic Materials and the Theory of International Trade*, Harvard University Press, Cambridge, MA, USA.

Hulse, S.H., Deese, J. and Egeth, H. (1975), *The Psychology of Learning* (4th ed.), McGraw Hill, USA.

Imai M. (1986), *KAIZEN: The Key to Japan's Competitive Success*, McGraw-Hill, New York, NY, USA.

Ishikawa, K. (1972), *Japan Quality Control Circles*, JSUE Press, Tokyo, Japan.

Jewell, W.S. (1984), "A generalized framework for learning curve reliability growth models", *Operations Research*, 32(3), 547-558.

Joskow, P.L. and Rozanski, G.A. (1979), "The effects of learning by doing on nuclear plant operating reliability", *Review of Economics and Statistics*, 61(2), 161-168.

Kaloo, U. and Towill, D.R. (1979), Time-dependent changes in the production levels of experienced workers", *Int. J. Production Research*, 17(9), 45-59.

Karger, D.W.K. and Bayah, F.H. (1977), *Engineered Work Measurement* (3rd ed.), Industrial Press Inc., New York, USA.

Keachie, L. and Fontana, J. (1966), "Effects of learning on optimal lot size", *Management Science*, 13, B102-B108.

Khoshnevis, B., Wolfe, P.M. and Terrell, M.P. (1982), "Aggregate planning models incorporating productivity - an overview", *Int. J. Production Research*, 20(5), 555-564.

Khoshnevis, B. and Wolfe, P.M. (1983a), "Aggregate production planning models incorporating dynamic productivity: Part I: model development", *IIE Trans.*, 15, 111-117.

Khoshnevis, B. and Wolfe, P.M. (1983b), "An aggregate production planning model incorporating dynamic productivity: Part II: solution methodology and analysis", *IIE Trans.*, 15, 283-291.

Khoshnevis, B. and Wolfe, P.M. (1986), "A short-cycle product aggregate planning model incorporating dynamic productivity", *Engineering Costs and Production Economics*, 10, 217-233.

Kiechel, W. (1981), "The decline of the experience curve", *Fortune*, Oct. 3, 139-145.

Kilbridge, M. (1959), "Predetermined learning curves for clerical operations", *J. of Industrial Engineering*, 10, 203-209.

Kilbridge, M. (1962), "A model for industrial learning costs", *Management Science*, 8(4), 516-527.

Kini, R.G. (1994), Economics of Conformance Quality. Unpublished PhD Dissertation, G.S.I.A. Carnegie Mellon University, PI, USA.

Klastorin, T.D. and Moinzadeh, K. (1989), "Production lot-sizing under learning effects: an efficient solution technique", *IIE Trans.*, 21, 2-10.

Klatzky, R.L. (1980), *Human Memory - Structures and Processes*, W.H. Freeman, San Francisco, CA. USA.

Klepper, R., Litecky, C. and Jones, W. (1989), "Self-managed teams and MIS productivity: literature, model and field study", *Data Base*, 20(1), 36-38.

Knecht, G.R. (1974), "Costing, technical growth and generalized learning curves", *Operational Research Quarterly*, 25(3), 487-491.

Konz, S. (1995), *Work Design: Industrial Ergonomics*, (4th ed.), John Wiley and Sons, NY.

Kopcso, D.W. and Nemitz, W.C. (1983), "Learning curves and lot sizing for independent and dependent demand", *J. Operations Management*, 4(1), 73-83.

Kottler, J.L. (1964), "The learning curve, a case history in its application", *The J. of Industrial Engineering*, 15(4), 176-180.

Kneip, J. (1965), "The maintenance progress function", *J. of Industrial Engineering*, 16(6), 398-400.

Kroll, D.E. and Kumar, K.R. (1989), "The incorporation of learning in production planning models", *Annals of Operations Research*, 17, 291-304.

Kvalseth, T.C. (1976), "The effect of practice on human performance in a preview constrained motor task", *J. of Motor Behavior*, 8, p. 143.

Kvalseth, T.C. (1978), "The effect of task complexity on the human learning function", *Int. J. Production Research*, 16(5), 427-435.

Kyllonen, P.C. and Alluisi, E.A. (1987), "Learning and forgetting facts and skills", in *Handbook of Human Factors*, Salvendy, G. (ed.), Wiley and Sons, New York, USA.

Lampton, D.R., Bliss, J.P. and Meet, M. (1992), The Effects of Differences Between Practice and Test Criteria on Transfer and Retention of a Simulated Tank Gunnery Task. Technical Report 949, US Army Research Institute for the Behavioral and Social Sciences, Alexandria, VA, USA.

Lanchman, R., Lanchman, J.L. and Butterfield, E.C. (1979), *Cognitive Psychology and Information Processing: An Introduction*, Wiley and Sons, New York, USA.

Lane, N.E. (1986), Skill Acquisition Curves and Military Training. IDA Paper P-1945, Institute for Defense, Alexandria, VA, USA.

Lehrer, R.N. (1957), *Work Simplification*, Prentice-Hall, Englewood Cliffs, NJ, USA.

Lesieur, F.G. (1958), *The Scanlon Plan*, The MIT Press, Cambridge, MA, USA.

Levin, N. and Globerson, S. (1993), "Generating learning curves for individual products from aggregated data", *Int. J. Production Research*, 31(12), 2807-2815.

Levitt, B. and March, G. (1988), "Organizational learning", *Annual Review of Sociology*, 14, 319-340.

Levy, F.K. (1965), "Adaptation in the production process", *Management Science*, 11(6), B134-B156.

Li, C.L. and Cheng, T.C.E. (1994), "An economic production quantity model with learning and forgetting considerations", *Production and Operations Management*, 3, 118-132.

Li, G. and Rajagopalan, S. (1997a), "The impact of quality on learning", *J. of Operations Management*, 15, 181-191.

Li, G. and Rajagopalan, S. (1997b), "A learning curve model with knowledge depreciation", *European J. of Operations Research*, 105(1), 143-154.

Li, G. and Rajagopalan, S. (1998), "Process improvement, quality and learning effects", *Management Science*, 44(11), 1517-1532.

Liao, W.M. (1979), "Effects of learning on resource allocation", *Decision Sciences*, 10(1), 116-125.

Lieberman, M.B. (1984), "The learning curve, pricing, and market structure in the chemical processing industries", *Rand J. Economics*, 15, 213-228.

Lippert, S. (1976), "Accounting for prior practice in skill acquisition studies", *Int. J. Production Research*, 14(2), 285-293.

Livine, S. and Melamed, A. (1991). Predicting the Learning Process for Long Tasks. Project term paper, Technion, Israel (in Hebrew).

Logan, G.D. (1988), "Toward an instant theory of automatization", *Psychological Review*, 95(4), 492-527.

Lundberg, E. (1961), *Produktivitet och Rantabilitet*, Norstedt and Soner, Stockholm, Sweden.

Magjuka, R.J. (1992), "Survey: self-managed teams achieve continuous improvement best", *National Productivity Review*, 11(1), 51-57.

Manivannan, S. and Hong, C.F. (1991), "A new heuristic algorithm for capacity planning in a manufacturing facility under learning 0", *Int. J. Production Research*, 29(7), 1437-1452.

Mansfield, E. (1961), "Technical change and the rate of imitation", *Econometrica*, 29, 741-766.

Maslow, A. (1970), *Motivation and Personality* (2nd ed.), Harper and Row, New York, NY, USA.

Mazur, J.E. and Hastie, R. (1978), "Learning as accumulations: a reexamination of the learning curve", *Psychological Bulletin*, 85(6), 1256-1274.

McCambell, E.W. and McQueen, C.W. (1956), "Cost estimating from the learning curve", *Aeronautical Digest*, 73, p. 36.

McCann, J. and Galbraith, J.R. (1981), "Interdepartmental relations", pp. 60-84, in *Handbook of Organizational Design* (Vol. 2), Nystrom, P.C. and Starbuck (eds.), Oxford University Press, New York, USA.

McDonald, J. (1987), "A new model for learning curves, DARM", *American Statistical Association*, 5(3), 329-338.

McGregor, D. (1960), *The Human Side of Enterprise*, McGraw-Hill, New York, NY, USA.

McIntyre, E.V. (1977), "Cost-volume-profit analysis adjusted for learning", *Management Science*, 24(2), 149-160.

McKenna, K., Glendon, S.P. and Glendon, A.I. (1985), "Decay in cardiopulmonary resuscitation skills", *J. Occupational Psychology*, 58, 109-117.

Middleton, K. (1945), "Wartime productivity changes in the airframe industry", *Monthly Labor Review*, 61, 215-225.

Miles, L.D. (1961), *Techniques of Value Analysis and Engineering*, McGraw-Hill, New York, NY, USA.

Minter, A.I. (1977), The Measure of Activity of Manual Work. Unpublished PhD dissertation, University of London, UK.

Mishina, K. (1987), "Behind the flying fortress learning curve", Term Project, Harvard Business School, May, '87.

Mogonson, A.H., (1963), "Work simplification - a program of continuing improvement", in *Industrial Engineering Handbook* (2nd edition), Maynard, H.B. (ed.), McGraw Hill, New York, NY, USA.

Monden, Y. (1983), *Toyota Production System*, Industrial Engineering and Management Press, Atlanta, GA, USA.

Montgomery, D.B. and Day, G.S. (1985), "Experience curves: evidence, empirical issues and applications", pp. 213-238, in *Strategic Marketing and Management*, Thomas, H. and Gardner, D. (eds.), Wiley and Sons, New York, NY, USA.

Moore, B.E. and Ross, T.L. (1978), *The Scanlon Way to Improved Productivity*, Wiley, New York, NY, USA.

Moravec, M., Johannessen, O.J. and Hjelmas, T.A. (1998), "The well-managed SMT", *Management Review*, 87(6), 56-58.

Morooka, K. and Nakai, S. (1971), Report by Committee of Learning Investigation to Japan Society of Mechanical Engineers, Feb. 1971.

Mudge, A.E. (1971), *Value Engineering: A Systematic Approach*, McGraw -Hill, New York, NY, USA.

Mundel, M.E. and Danner, D.L. (1995), *Motion and Time Study* (7th edition), Prentice-Hall, Englewood Cliffs, NJ, USA.

Muth, E.J. and Spremann, K. (1983), "Learning effects in economic lot sizing" *Management Science*, 29(2), 264-269.

Nadler, G. and Smith, W. (1963), "Maufacturing progress functions for types of processes", *Int. J. Production Research*, 2(2), 115-135.

Nadler, G. (1967), *Work Systems Design: The IDEALS Concept*, Irwin Pub. Co., Homewood, IL. USA.

Nadler, G. (1970), *Work Design: A Systems Concept*, Irwin Pub. Co., Homewood, IL, USA.

Nadler, G. (1981), *The Planning and Design Approach*, Wiley and Sons, New York, NY, USA.

Nadler, G. and Hibino, S. (1998), *Breakthrough Thinking* (2nd edition), Prima Pub. Co., Rocklin, CA, USA.

Naim, M.M. and Towill, D.R. (1993), "Modeling and forecasting industrial innovations via the transfer function S-shaped learning curve", *Int. J. Advanced Manufacturing Technology*, 8, 329-343.

Nanda, R. and Adler, G.L. (1977), *Learning Curves, Theory and Application*, WS and ME Monograph 6, American Institute of Industrial Engineers, Norcross, GA, USA.

Nander, R. and Nam, H.K. (1992), "Quantity discounts using a joint lot size model under learning effects - single buyer case", *Computers and Industrial Engineering*, 22, 211-221.

Nander, R. and Nam, H.K. (1993), "Quantity discounts using a joint lot size model under learning effects - multiple buyer case", *Computers and Industrial Engineering*, 24, 487-494.

Neibel, B.W. (1976), *Motion and Time Study* (6th edition), Richard D. Irwin, Homewood, IL, USA.

Nelson, R.R. and Langlois, R.N. (1983), "Industrial innovation policy: lessons from American history", *Science*, 219, 814-819.

Nevis, E.C., DiBella, A.J. and Gould, J.M. (1995), "Understanding organizations as learning systems", *Sloan Management Review*, 36, 73-85.

Newell, A. and Rosenbloom, P.S. (1981), "Mechanisms of skill acquisition and the law of practice", pp. 1-55 in *Cognitive Skills and their Acquisition*, Anderson, J.R. (ed.), Lawrence Erlbaum Associates, Hillsdale, NJ, USA.

Noyce, R. (1977), "Microelectronics", *Scientific America*, 237(3), 63-69.

O'Hara, M.J. (1988), "The retention of skills acquired through simulation-based training", *Ergonomics*, 33(9), 1143-1153.

Omachonu, V.K. and Ross, J.E. (1994), *Principles of Total Quality*, St. Luce Press, Delray Beach, FL, USA.

Pachella, R. (1974), "The use of Reaction Time Measures in Information Processing Research", in *Human Information Processing*, Kantowitz, B.H. (ed.), Earlbaum Associates, Hillsdale, NJ, USA.

Pegels, C.C. (1969), "On startup or learning curves: an expanded view", *AIIE Trans.*, 1(3), 216-222.

Pegels, C.C. (1976), "Start up or learning curves - some new approaches", *Decision Sciences*, 7(4), 705-713.

Person, H.B. (1989), "Review of analytic and simulation studies of Just-in-Time systems with Kanbans", pp. 984-986, in the *Proceedings of the 20th Annual Meeting of the Decision Sciences Institute*, 20-22 November, USA.

Pew, R.W. (1969), "The speed-accuracy operating characteristic", *Acta Psychologia*, 30, 16-26.

Pratsini, E., Camm, J.D. and Raturi, A.S. (1993), "Effects of process learning on manufacturing schedules", *Computers and Operations Research*, 20(1), 15-24.

Prophet, E. (1976), Long-Term Retention of Flying Skills: A review of the Literature, HUMRRO/FR-EDP-76-35. Human Resources Research Organization, Alexandria, VA. USA.

Rabinovitch, M. (1983), Determining the Optimal Cycle Time in Assembly Lines. Unpublished MSc Thesis, Technion, Israel (in Hebrew).

Rao, A., Carr, L.P., Dambolena, I., Kopp, R.J., Martin, J., Rafii, F. and Schlesinger, P.F. (1996), *Total Quality Management: A Cross Functional Perspective*, J. Wiley and Sons, NY, USA.

Rapping, L. (1965), "Learning and the World War II production functions", *Review of Economics and Statistics*, 48, 98-112.

Richardson, W.J. (1978), "Use of learning curves to set goals and monitor progress in cost reduction programs", pp. 235-239, in *Proceedings of the 1979 IIE Spring Conference*.

Roberts, P. (1983), "A theory of the learning process", *J. of the Operations Research Society*, 3, 71-79.

Rogers, H. (1998), "Benchmarking your plant against TQM best-practice plants", *Quality Progress*, 31(3), 49-55.

Rosenwasser, C.E. (1982). The Influence of Job Complexity on Learning Curves. Unpublished MSc Thesis, Technion, Israel.

Rousseau, D.M. (1997), "Organizational behavior in the new organization", *Annual Review of Psychology*, 48, 515-546.

Rucker, A.W. (1962), Gearing Wages to Productivity. The Eddie-Rucker-Nickels Co, USA.

Ruffner, J.W. and Bickley, W. (1984), Retention of Helicopter flight skills: Is there a 'critical period' for proficiency loss?, pp. 370-374, in *Proceedings of the Human Factors Society, 28th Annual Meeting*, Santa Monica, CA, USA.

Sabag, K. (1988), Prediction of Learning in Long Cycle Times. Unpublished MSc Thesis, Technion, Israel (in Hebrew).

Sagoe, I.K. (1994), "Why are self-managed teams so popular?", *J. for Quality and Participation*, 17(5), 64-67.

Sahal, D. (1979), "A theory of progress functions", *AIIE Trans.*, 11(1), 23-39.

Salavendy, G. and Pilitsis, J. (1974), "Improvements in physiological performance as a function of practice", *Int. J. of Production Research*, 12(4), 519-531.

Salomone, T.A. (1995), *What Every Engineer Should Know About Concurrent Engineering*, Dekker, New York, NY, USA.

Schendel, J.D., Shields, J.L. and Katz, M.S. (1975), Retention of Motor Skills: Review. Technical Report 313, Army Research Institute, Alexandria, VA. USA.

Schmenner, R.W. and Cook, R.L. (1985), "Explaining productivity differences in North Carolina factories", *J. of Operations Management*, 5(5), 273-289.

Schneider, W. (1989), "Training models to estimate training costs for new systems", Chapter 17, pp. 215-232, in *Human Performance Models for Computer-Aided Engineering*, Elkind, J.I., Card, S.K., Hochberg, J. and Huey, B.M. (eds.), National Academy Press, Washington DC.

Schonberger, R.J. (1982), *Japanese Manufacturing Techniques, Nine Hidden Lessons in Simplicity*, The Free Press, New York, USA.

Schurmans, D. (1997), "Characterizing rational versus exponential learning curves", *J. of Computer and System Sciences*, 55, 140-160.

Searle, A.D. and Goody, C.S. (1945), "Productivity increases in selected wartime shipbuilding programs", *Monthly Labor Review*, 61, 1132-1147.

Semple, C.A., Hennesy, R.T., Sanders, M.S., Cross, B.K., Beith, B.H. and McCauley, M.F. (1981), Aircrew Training Devices: Fidelity Features. Logistics and Technical Training Division, Wright Patterson Air Force Base, OH, USA.

Senge, P.M. (1990), *The Fifth Discipline: The Art and Practice of the Learning Organization*, Doubleday/Currency, New York, NY, USA.

Sexton, C. (1994), "Self-managed work teams: TQM technology at the employee level", *J. of Organizational Change Management*, 7(2), 45-52.

Sharp, J.A. and Price, D.H.R. (1990), "Experience curves in the electric supply industry", *Int. J. of Forecasting*, 6(4), 531-540.

Sheshinski, E. (1967), "Tests of the learning by doing hypothesis", *Review of Economics and Statistics*, 49, 568-578.

Shewhart, W.A. (1939), *Economic Control of Quality of Manufacturing Product*, Van Nostrand Reinhold, New York, NY, USA.

Shields, J.L., Goldberg, S.L. and Dressel, J.D. (1979), Retention Basic Soldiering Skills. Army Research Report 1225, Army Research Institute, Alexandria, VA. USA.

Shtub, A, (1991), "Scheduling of programs with repetitive projects", *Project Management Journal*, 22(4), 49-53.

Shtub, A, Levin, N. and Globerson, S. (1993), "Learning and forgetting industrial skills: an experimental model", *Int. J. Human Factors in Manufacturing*, 3(3), 293-305.

Sims, H.P. and Dean, J.W. (1985), "Beyond quality circles: self-managing teams", *Personnel*, 62(1), 25-32.

Skinner, W. (1969), "Manufacturing - missing link in corporate strategy", *Harvard Business Review*, 47(3), 136-145.

Smeler, R. (1993), *Maverick: The Success Story Behind the World's Most Unusual Workplace*, Warner Books, NY, USA.

Smith, J. (1989), *Learning Curve for Cost Control*, Industrial Management Press, Norcross, GA, USA.

Smith, P.A. and Smith, K.V. (1955), "Effects of sustained performance on human motions", *Perpetual Motor Skills*, 5, 23-29.

Smunt, T.L. and Morton, T.E. (1985), "The effects of learning on optimal lot sizes: further development on the single product case", *IIE Trans.*, 17, 33-37.

Smunt, T.L. (1986a), "Incorporating learning curve analysis into median-term capacity planning procedures: a simulation experiment", *Management Science*, 32(9), 1164-1176.

Smunt, T. L. (1986b), "A comparison of learning curve analysis and moving average ratio analysis for detailed operational planning", *Decision Sciences*, 17(4), 475-495.

Smunt, T.L. (1987), "The impact of worker forgetting on production scheduling", *Int. J. Production Research*, 25(5), 689-701.

Smunt, T.L. (1996), "Rough cut capacity planning in a learning environment", *IEEE Trans. on Engineering Management*, 43(3), 334-341.

Snoddy, G.S. (1926), "Learning and stability", *J. of Applied Psychology*, 10, 1-36.

Sparks, C. and Yearout, R. (1990), "The impact of visual display units used for highly cognitive tasks on learning curve models", *Computers and Industrial Engineering*, 19(4), 351-355.

Spence, A.M. (1981), "The learning curve and competition", *Bell J. of Economics*, 13(1), 20-35.

Spencer, R.J. (1995), "Success with self-managed teams and partnering", *J. for Quality and Participation*, 18(4), 48-53.

Spradlin, B. and Pierce, D. (1967), "Production scheduling under a learning effect by dynamic programming", *J. of Industrial Engineering*, 18, 219-222.

Sriyananda, H. and Towill, D.W. (1973), "Prediction of human operator performance", *IEEE Trans.*, R-22, 148-158.

Stapper, C.H. (1989), "Fact and fiction on yield modeling", *Microelectronics*, 20(1), 129-151.

Stata, R. (1989), "Organizational learning: the key to management innovation", *Sloan Management Review*, 30(3), 64-74.

Steedman, I. (1970), "Some improvement curve theory", *Int. J. Production Research*, 8(3), 189-205.

Sule, D.R. (1978), "The effect of alternate periods of learning and forgetting on economic manufacturing quantity", *AIIE Trans.*, 10(3), 338-343.

Sule, D.R. (1981), "A note on production time variation in determining EMQ under influence of learning and forgetting", *AIIE Trans.*, 13(1), 91.

Sule, D.R. (1983), "Effect of learning and forgetting on economic lot size scheduling problem", *Int. J. Production Research*, 21(5), p. 771.

Swenseth, S.R., Muralidhar, K. and Wilson, R.L. (1993), "Planning for continual improvement in a Just-in-Time environment", *Int. J. of Operations and Production Management*, 13(6), 4-22.

Tanskanen, T., Buharnist, P. and Kostama, H. (1998), "Exploring the diversity of teams", *Int. J. of Production Economics*, 57, 611-619.

Thomopoulos, N.T. and Lehman, M. (1969), "The mixed-model learning curve", *AIIE Trans.*, 1, 127-132.

Tinnin, D.B. (1983), "How IBM stung Hitachi", *Fortune*, March 7, 50-56.

Titleman, M.S. (1957), "Learning curves - key to better labor estimates", *Product Engineering*, 29, p. 36.

Tjosvold, D. (1991), *Team Organization: An Enduring Competitive Advantage*, Wiley and Sons, New York, NY, USA.

Towill, D.R. and Bevis, F. (1973), "Managerial control systems based on learning curve models", *Int. J. Production Research*, 11(3), 219-238.

Towill, D.W. (1982), "How complex a learning curve need we use?", *Radio and Electronic Engineer*, 52(7), 331-338.

Towill, D.W., Davies, A. and Naim, M.M. (1989), "The dynamics of capacity planning for flexible manufacturing system startup", *Engineering Costs and Production Economics*, 17, 55-64.

Towill, D.R. (1989), "Selecting learning curve models for human operator performance", pp. 403-417, in *Applications of Human Performance Models to Systems Design*, Plenum Press, UK.

Towill, D.R. (1990), "Forecasting learning curves", *Int. J. of Forecasting*, 6, 25-38.

Towill, D.R. and Cherrington, J.E. (1994), "Learning curve models for predicting the performance of AMT", *Int. J. Advanced Manufacturing Technology*, 9, 195-203.

Turban, E. (1968), "Incentives during learning, an application of the learning curve theory on a survey and other methods", *The J. of Industrial Engineering*, 19(12), 600-607.

Uezu, T. and Kabashima, Y. (1996), "Perfect learning and power law in learning from stochastic examples by Ising perceptions: analysis under one-step replica symmetry breaking ansatz", *J. Physics A: Math. Gen.*, 29, L439-L445.

Underwood, B.J. (1954), "Speed of learning and amount retained: a consideration of methodology", *Psychological Bulletin*, 51(3), 276-282.

Underwood, B.J. (1968), "Forgetting", pp. 536-542 in the *International Encyclopedia of the Social Sciences*, Vol. 5, Stills, D.L. (ed.)., Macmillan, New York, USA.

Van Cott, H.P. and Kinkade, R.G. (eds.) (1972), *Human Engineering Guide to Equipment Design*, American Institute for Research, Washington, DC, USA.

Venda, V. (1995), "Ergodynamics: theory and application", *Ergonomics*, 38, 1600-1616.

Vollichman, (1993), Speed versus Accuracy Under Continuous and Intermittent Learning. Unpublished MSc Thesis, Technion, Israel (in Hebrew).

Von Hippel, E. (1976), "The dominant role of users in the scientific instruments innovation process", *Research Policy*, 5(3), 212-239.

Von Tetra, P. and Smith, K.V. (1952), "The dimensional analysis of motions: IV. Transfer effects and direction of movement", *J. of Applied Psychology*, 36, 348-353.

Ward, J.A. (1997), "Implementing employee empowerment", *Information Systems Management*, 14(1), 62-65.

Werner International - Textile Management Consultants, New York, Brussels (1972).

Wheelwright, S.C. and Clark, K.B. (1992), *Revolutionizing Product Development*, The Free Press, New York, NY, USA.

Wickelgren, W.A. (1977), *Learning and Memory*, Prentice-Hall, Englewood Cliffs, NJ, USA.

Wickelgren, W.A. (1981), "Human learning and memory", *Ann. Rev. Psychology*, 32, 21-52.

Wickens, C.D.(1984), *Engineering Psychology and Human Performance*, Charles E. Merrill Publishing Co., Columbus, OH, USA.

Wilhem, A.T. and Sastri, T. (1979), "An investigation of flow line operating characteristics during start-up", *Int. J. Production Research*, 17(4), 345-358.

Wisher, R., Sabol, M. Sukenik, H. and Kern, R. (1991), Individual Ready Reserve (IRR) call up: Skill decay. Research Report 1595, US Army Research Institute, Alexandria, VA, USA.

Womer, N.K. (1979), "Learning curves, production rate, and program costs", *Management Science*, 25(4), 312-319.

Womer, N.K. and Patterson, J.W. (1983), "Estimation and testing of learning curves", *J. of Business Economics and Statistics*, 1, 265-272.

INDEX

234

236